# Problem Solving with Creative Mathematics

George Bradley
Duquesne University

Brooks/Cole Publishing Company

I(T)P An International Thomson Publishing Company

Pacific Grove • Albany • Bonn • Boston • Cincinnati • Detroit • London • Madrid • Melbourne
Mexico City • New York • Paris • San Francisco • Singapore • Tokyo • Toronto • Washington

A GARY W. OSTEDT BOOK

Sponsoring Editor: Gary W. Ostedt
Marketing Representative: Dane McKaughan
Editorial Associate: Carol Ann Benedict
Production Editor: Tessa A. McGlasson
Cover Design: Roy R. Neuhaus
Art Coordinator: Lisa Torri
Printing and Binding: Malloy Lithographing, Inc.



*For more information, contact:*

Brooks/Cole Publishing Company
511 Forest Lodge Road
Pacific Grove, CA 93950
USA

International Thomson Editores
Campos Eliseos 385, Piso 7
Col. Polanco
11560 México D. F. México

International Thomson Publishing—Europe
Berkshire House 168-173
High Holborn
London WC1V 7AA
England

International Thomson Publishing GmbH
Königwinterer Strasse 418
53227 Bonn
Germany

Thomas Nelson Australia
102 Dodds Street
South Melbourne, 3205
Victoria, Australia

International Thomson Publishing—Asia
221 Henderson Road #05-10
Henderson Building
Singapore 0315

Nelson Canada
1120 Birchmount Road
Scarborough, Ontario
Canada M1K 5G4

International Thomson Publishing—Japan
Hirakawacho-cho Kyowa Building, 3F
2-2-1 Hirakawacho-cho
Chiyoda-ku, Tokyo 102
Japan



Printed in the United States of America.

5 4 3 2 1

ISBN 0-534-24360-6

THIS BOOK IS PRINTED ON ACID-FREE RECYCLED PAPER

To my parents, Martha and Russell Bradley

For all their love and support throughout my 40 years.

## Foreword

When the University Core Curriculum was being discussed at Duquesne University, it was agreed that a mathematics course should be part of every undergraduate's university experience. The Department of Mathematics and Computer Science decided that a course in problem solving would be far more beneficial to students not requiring mathematics for their majors than a course in college algebra. Ideally, we would have liked to require every undergraduate student at Duquesne to take a course in problem solving; however, some schools in the University, such as Pharmacy, Business and Nursing, already required mathematics courses for their students and could not accommodate another required course. Therefore, it was decided that any undergraduate student not taking calculus or statistics would be required to take our course in problem solving. Our course was initially developed by Dr. Kathleen Taylor, Professor of Mathematics at Duquesne University, and pilot-tested by her. Money provided by the Pew Trusts was used to fund a week-long workshop where Mark Mazur, Charles Loch, the late Rosaline Lee, Kathy Taylor and I further refined the course. Our course has been taught since 1987.

## To the Instructor

You might be using this book as either a supplement to Karl Smith's text or similar book, or as a core text together with material of your choosing. If this is being used as a supplementary text, you may use the exercises to introduce the students to a topic instead of introducing it through lecture. Most of the exercises in this book have been used in our course at Duquesne. They have been designed as class exercises, but many of them could be assigned to be completed outside of class. Ample space is left after each question for students to give complete answers. The questions should be answered as they are encountered. The exercise may then be removed from the book and handed in to you. More detailed information may be provided on that topic in a lecture, and then homework problems from the text may also be assigned.

Students in our course participate much more actively in the learning process than they would in a traditional class. Class time is spent having the students complete the exercises in the book or discuss problems such as those that appear at the end of many of the chapters or in the appendices at the end of the text. We do not lecture, although occasionally we will involve the entire class in a discussion. Students work on the exercises individually or in small groups of four or five. The other problems are much more complex and are discussed by the students in the small group setting. We generally give the students fives sets of such problems each semester, with four or five problems on each set. What we are looking for and how the students are evaluated on these problems is discussed more fully under the section "To the Student." In our course, grades are determined as follows: 20% class participation, 15% for the midterm examination, 15% for the final examination, and 50% based on the exercises, problem sets, and other activities of the faculty member's choosing, many examples of which are given in this book. Guidelines for small group dynamics, sample midterm and final exams, a large list of problems, samples of what we expect for solutions to the problems, and other information giving more details about our course are available free upon request from the publisher.

Please send any comments about this book to George Bradley, Dept. of Mathematics and Computer Science, Duquesne University, Pittsburgh, PA 15282.

## To the Student

This book was specifically designed to help you "develop an appreciation of mathematics as an intellectual discipline and as a problem solving tool." Although it is widely accepted that mathematics is necessary for modern technology, it is not only a tool for the sciences but makes a basic contribution to all aspects of human life. Problem solving, done by yourself and with others, is important in every discipline. Mathematics is an art as well as a science; it has structure and beauty just as do the visual and musical arts. Mathematics is inflluenced by and affects our culture and history. You will see how to integrate your knowledge of and study of mathematics with other experiences and disciplines. Mathematics is far more wide-ranging than you probably ever realized. And discovering some basic mathematical principles on your own is fun!

Mathematics is a language; it is the language of quantity. It is logically structured; its study helps to introduce you to abstract and analytical thinking. Its focus is understanding concepts, not memorizing algorithms. Anyone who can read and is willing to work can improve her/his "Problem Solving" ability. Doing the exercises in this book will aid you in learning to read, interpret, and solve mathematical problems; just as importantly, it will also aid you in learning to write and communicate mathematical ideas so that others can understand. You will discover basic mathematical principles on your own, which will lead to a better understanding of the material and also give you some confidence in solving mathematical problems. You will generalize a solution to a specific problem so that solutions to an entire class of related problems may be found. You will develop the ability to see the application of mathematical principles in several areas. You will examine problems and their feasible solutions from many different points of view, learn strategies for solving problems, and realize that you <u>can do and enjoy</u> mathematics!

The solution to a problem for group discussion involves far more than a single number, word, or even sentence. The following guidelines should be followed when you are writing up problems. Your instructor may add to or modify them.

Form: Each problem and solution should begin on a new sheet of paper. State the problem and then begin the solution. Your problem set may be typed or hand written if your handwriting is legible. You are expected to use clear, standard English, correct sentence structure and grammar, and correct spelling. Do not hand in work which contains crossed out errors or torn pages or papers with ragged edges.

In grading your problem sets, your instructor is looking for:

1) Coherent and clear explanations of how the answer was obtained (this includes neatness).

2) Correct answers.

3) Analysis of the solution:

   a) Is this the only possible correct answer? Are there other ways of getting the same answer? Is one solution or method better than the others?

   b) How would changing the conditions of the problem change the solution?

   c) Is this problem related to others you have seen in this class or in other situations outside of class?

   d) What general principles, if any, have you deduced from this problem?

The assessment of your work will necessarily be more subjective than you are used to in mathematics classes. The above considerations will be weighed approximately as follows (your instructor may give you further or different guidelines):

70% in categories 1) and 2).
30% in category 3).

When completing the exercises in this book, you will see that ample space has been left for you to write your answers to the posed questions. Please make sure you write the answers using complete sentences, and, when appropriate, complete paragraphs.

## Acknowledgments

I would like to thank Dane McKaughan, the former Brooks/Cole representative for our campus, who encouraged me to write the prospectus for this book and was instrumental in getting this project started. Special thanks go to Gary Ostedt for his interest, faith and help with this project, and to his assistant, Carol Benedict. Thanks to all of the reviewers whose comments were very valuable. Thanks to Ruth Kittner, who read the completed manuscript and made many valuable suggestions, and to Jim McKelvey, who also made helpful comments when reviewing the manuscript in its preliminary stages. Thanks to Marguerite Puhl, my "niece-in-law" Melissa Batt, and my father Russ Bradley for their artwork for the book. Thanks to my mother, Martha Bradley, my sister Pam, her husband Bruce and children Mike and Laura for all their support. Thanks to Tom Keagy, our department chair, for his support and valuable suggestions, and to Dan Barbush, the members of our department mentioned earlier who helped to develop our course, and the rest of our department for all they have done. Finally, thanks to all my friends, old and new, who have been there to encourage me throughout my work on this project.

# TABLE OF CONTENTS

# CHAPTER ONE - BASIC PROBLEM SOLVING

Almost anyone can learn to read, but few become great writers. Many people can learn to cook, but few become great chefs. Many can learn to swing a baseball bat and catch a ball, but few become great baseball players. Many learn to play a musical instrument, but few become professional musicians. Still, many of us enjoy reading, cooking, music, and playing sports. In the same way, almost anyone can learn and understand many basic mathematical principles, even though few people become great mathematicians. And you don't have to be a mathematician to enjoy mathematics!

Many students (including me, the author) have had bad experiences with mathematics in the past. Often, the problem is just with one concept or one teacher, but that experience can cause students to give up, thinking that mathematics is beyond them. You may think that you are alone in your dislike, love, fear, hatred, or whatever feeling you have toward mathematics. Try the following exercises and you'll change your mind.

Do the first exercise in class on the first day, and hand in the second exercise on the second class day.

1. Write one or two paragraphs on your feeling about taking this course and about mathematics in general. Be honest! After your paragraphs have been turned in, have your professor read some of them (you will remain anonymous!).

2. Write a one page mathematical autobiography. Talk about your experiences with mathematics over the years. Have you always enjoyed mathematics? Always disliked it, even in first grade? If neither of these, talk about when your feelings began to change. Have your feelings about mathematics influenced your career goals? After you have completed this exercise, ask your professor to read (anonymously) some of the comments made.

Many students have lost the interest and confidence they once had in tackling mathematical problems. To be able to solve problems takes patience, some time to study the problem, and a strategy to follow, not genius.

## Problem Solving Strategies (many are adapted from George Polya's How to Solve It).

1. What do you know? What are you looking for?

   a. Read the problem over and over again.
   b. Restate the problem in your own words.
   c. Draw a picture.
   d. Make a chart.
   e. Long periods of sensible exploration may be needed.

2. Plotting the way to a solution.

   a. Trial and error.
   b. Could you answer a simpler related problem?
   c. Examine special cases; try the problem for different small integer values.
   d. Look at a series of examples to see if a pattern emerges.
   e. Drudgework may be necessary.
   f. Go back to the conditions of the problem, explore them, and look for connections between them and the desired quantities.
   g. If there are a large number of variables, all of which play the same role, look at the problem with only one or two of the variables.
   h. Working backwards may help.

3. Solve the problem.

4. Look at the solution.

   a. Does your solution make sense?
   b. Is yours the only solution possible? Is there another solution?
   c. Can your solution be generalized or is it a particularization of a known principle?
   d. Have you seen other problems related to this one? How are they similar? How are they different?
   e. Can you make up problems similar to this one by changing one or two conditions? How is the solution changed?

Whether you are doing one of the introductory problems, a problem at the end of a chapter or at the end of the text please keep these suggestions in mind:

1. State the problem before solving it. That way, your write-up will be self contained.
2. Do not use paper with ragged edges.
3. Start each problem on a new page.
4. Be very complete in your explanation. You should pretend that the person reading your solution has not seen this problem before and has no idea of how to solve it.
5. Check your solution and its explanation to make certain that they are correct.
6. Use correct spelling, complete sentences, and all standards of good writing when writing up the solution.
7. Realize that simply working examples does not justify the correctness of a statement. I could ask the question, "are all positive integers divisible by 2?" If I just considered a few examples, say 4, 12, 22, and 16, I might erroneously conclude that the answer is "Yes."

Here are some problems to get you rolling. These can be discussed in small groups in your class.

1. Twenty-five students are in a mathematics class. To get acquainted, each student shakes hands with each of the other students once and only once. What is the total number of handshakes that take place?

See if you can come up with a solution to this problem and justify whether you are correct.

Try to solve the problem for a small number of students to see if what you thought about the problem originally is correct. You can try acting it out with three, four, and five students. How many handshakes are there with these numbers of students? Solve the problem using two different methods. You will be able to derive a formula for finding the sum of consecutive integers.

2. Three boxes are labeled "CD's", "Cassettes", and "CD's and Cassettes." Each box is labeled incorrectly. You want to determine the correct labeling of the boxes, but you are allowed to select only one item from only one of the boxes. How can this be done?

Just choose one of the boxes and see what you can discover if a CD is chosen at random, and then if a cassette is chosen at random. In both cases you must be able to answer the question. Once you have discovered how it can be done, try looking at the problem to see if you can answer it if you do not know that all the boxes are labeled incorrectly, or if you have more boxes.

3. The planet Remulac uses slivers as its currency. The symbol for slivers is @. Suppose, to save paper, Remulac prints bills in only @5 and @7 denominations. Can Mr. Conehead buy @33 worth of popcorn with these denominations and receive exact change (possibly change could be zero)? What about @43 worth?

You can make your own currency if you want to. Try to see if these purchases can be made in more than one way. Also try to make exact purchases of other amounts. Can you find any that cannot be made?

Try answering the questions again with denominations such as @4 and @6 for the bills, and @6 and @9 for the bills. Write about any conclusions that you can make.

4. Suppose you have only two containers but an unlimited supply of water. The containers you have are for 5 liters and 3 liters. How can you measure exactly 4 liters of water if neither container has any measuring marks on it?

Can you find a second method? Which one is more efficient in terms of time or water used?
What other quantities of water can be measured?

5. From 1975 to 1985, the population of Mercury City increased by 20%. From 1985 to 1995, the population decreased by 20%. Would we be correct in saying that the population in 1975 and the population in 1995 must be the same?

6. Two players, Adrienne and Sam, take turns playing the following game. There is a pile of six toothpicks. At his or her turn, the player must take one or two toothpicks from those remaining in the pile. The player who takes the last toothpick wins. Adrienne must go first. If each player plays as intelligently as possible, who will win the game?
Would this be true for other numbers of toothpicks also?

7. How many squares are there on an ordinary checkerboard?

8. This problem was taken from a Greek text dating from the 500's. How many apples are there in a collection if the apples are distributed among six people in the following manner: the first receives 1/3 of the apples, the second 1/4, the third 1/5, the fourth 1/8, the fifth 10 apples, and the sixth 1 apple?

9. The following is a summary of the life of Diophantus of Alexandria, a mathematician of Ancient Greece: his boyhood wasted one-sixth of his life; his beard grew after one-twelfth more; he married after another seventh; his son was born five years later and lived to only half his father's age; and Diophantus died five years after his son.

How long did Diophantus live?

10. Analyze the game of Tic-Tac-Toe. What is the best strategy to pursue if you are player number 1. What is the best strategy if you are player number 2? Who should win the game?

# CHAPTER TWO - CONSUMER MATHEMATICS

A knowledge of some simple mathematics will help you to be a smarter consumer. Searching for the best buys, understanding discounts, etc., will certainly help you save money. The following exercises may help you learn to stretch your money a bit further.

## Exercise 1 - Coupons and Best Buys

Often you can purchase the same item in several different sizes. Assuming you do not have to worry about waste, it is always good to figure out which one is the best buy.

Rice Krinkles come in three sizes. A 10 oz. box sells for \$2.50, a 16 oz. box for \$3.20, and a 20 oz. box for \$4.20.

Which one is the best buy? Why?

Which is the worst buy? Why?

How do you think using coupons affects which size will give you the best buy? Do you think the same size will still be the best buy?

Decide which box would be the best buy and which would be the worst buy if you have a 25 cents off coupon.

What if you have a 50 cents off coupon?

A 75 cents off coupon?

Some stores double all coupons 99 cents and below. If your store does this, decide which is the best buy and which is the worst buy with a 25 cents, a 50 cents, and finally a 75 cents off coupon.

## Exercise 2 - Discounts

Suppose you decide to buy a new Paris Mist coat at Racy's department store. The price marked is $200, but all the coats are on sale for 20% off the marked price. How much do you actually have to pay for the coat? Explain.

Now answer the question if the marked price is P dollars and the discount is r% off the marked price.

The original price of a pair of cross-training Ekin shoes was $90. You were lucky enough to find the shoes for $63. What discount rate were you given? Explain.

Now answer the question if the original price was P dollars and the final price was S dollars.

Your friend Dan just bought a new color television set for $468. The manager of the Barn appliance store where he bought it told him he was getting a 22% discount off the marked price. If this is so, what was the marked price of the TV?

Now answer the question if the price Dan paid for the TV was S dollars and the discount was r%.

Sometimes discounts are added on top of other discounts, and you may expect to get more of a discount than you actually do. Suppose Bloominghill's department store is selling a set of new china originally marked for $100. The price was then reduced by 20%. But Bloominghill's is also offering an additional 30% on anything purchased during its President's Day Sale. How much money will that new set of china actually cost you?

What percentage of the original price are you going to pay?

How does that compare to the sum of 20% and 30%?

Suppose the china is marked 30% off the original price and then the store is giving an additional 20% off for all purchases. Now how much would you pay?

Is it better to get 30% off first and then 20% off, or 20% off and then 30%?

Now consider the problem again with discounts of first 40%, then 25%, and first 25%, then 40%. What conclusions can you make?

## Exercise 3 - Utilities

Many utilities, such as the gas and electric companies, charge their customers a monthly connection fee, in addition to the charge per unit used.

What would a customer's total electric bill for a month be if the monthly connection fee is $8.00, the cost of the electricity per kilowatt hour is 12 cents, and the number of kilowatt hours of electricity used is 900?

Here are a customer's electric bills for three consecutive months. Can you figure out what the connection fee is and how much each kilowatt hour of electricity is costing the consumer?

| | | |
|---|---|---|
| May | 700 kwh | $84.00 |
| June | 800 kwh | $95.00 |
| July | 760 kwh | $90.60 |

## Exercise 4 - Simple Interest

When simple interest is used for interest received on an investment, the interest earned each time period is computed only on the amount of the investment. When interest income is calculated, the interest paid each time period is based only on the original amount invested.

Let's consider investments made at simple interest. For example, if an investment is made at a simple interest rate of 4% per year, each year 4% of $1000, or $40, is earned in interest.

Consider this table of how much you would have at the end of each year if you invest $1000 today at a simple interest rate of 4% per year.

| Year | Balance |
|---|---|
| 0 | $1000 |
| 1 | $1040 |
| 2 | $1080 |
| 3 | $1120 |
| 4 | ______ |
| 5 | ______ |

Fill in the rest of the table. Can you describe in words how you determine how much interest you will have earned and what your total balance will be after n years?

Can you derive formulas for the amount of interest earned after n years and the total in the account after n years?

Suppose you invest P dollars at a simple interest rate of r per year for n years. How much money will you have then? (Find the formula).

## Exercise 5 - Compound Interest

When interest is compounded, interest is paid not only on the amount invested but on any previously accumulated interest as well. Do you think that you'll earn more than when simple interest was used, provided the interest rates are the same?

Suppose now you invest \$1000 at a simple interest rate of 4% compounded annually. That means every year you earn, as interest, 4% of whatever is in your account at that time.

Here's a table giving your balance:

| Year | Balance |
|---|---|
| 0 | \$1000 |
| 1 | \$1040 |
| 2 | \$1081.60 |
| 3 | ______ |
| 4 | ______ |

Fill in the rest of the table. Now try to find a method which will tell you how much money you will have in n years. The following calculations should help you.

After 1 year, $1000 + 1000(0.04) = 1000(1.04) = \$1040$

After 2 years, $1040 + 1040(0.04) = 1040(1.04) = \$1000(1.04)^2$ $= \$1081.60$.

After 3 years, $1081.60 + 1081.60(0.04) = 1081.60(1.04)$ $= 1040(1.04)^2 = 1000(\_\_\_\_\_\_) = \_\_\_\_\_\_\_\_$

How would you calculate how much money you will have after n years if you invest $1000 today at 4% annual interest, compounded annually?

If interest is compounded more frequently, then interest is awarded more frequently. For example, if interest is compounded twice per year (semiannually), then the interest is awarded twice per year. To determine how much interest is given, the annual rate must also be divided by two. So, if the interest rate is 4% per year, 2% of the current balance is given as interest.

Suppose $1000 is invested at an annual interest rate of 4%. Please complete the following table:

| Month | Balance. |
|---|---|
| 0 | $1000 |
| 6 | $1020 |
| 12 | $1040.40 |
| 18 | $1061.21 |
| 24 | ________ |
| 30 | ________ |
| 36 | ________ |

After 0.5 years, you have $1000(1.02) = \$1020$

After 1 year, you have $1020(1.02) = \$1000(1.02)^2 = \$1040.40$

How much will you have after 1.5 years?

After 2 years?

After n years?

Can you derive a general formula which will tell you how much money you will have in n years if you invest P dollars at interest rate r, compounded semiannually?

Try an investment of $1000 at 4% annual interest compounded quarterly (4 times per year). Make a table and try to derive some general formulas.

## Exercise 6 - Optimal Strategy for Selling a Car

Cher wishes to sell her old car. She places an ad in the newspaper and asks interested parties to make her an offer. She will have to decide which offer to take as the calls come in. She knows of three methods she could use to make this decision.

A. Accept the first offer she receives.

B. Accept the first offer that is better than the one just preceding.

C. Continue to consider offers until she gets one that is better than all those preceding. Accept that one.

In B or C, if none of the first three offers is accepted, accept the fourth offer.

If offers come to Cher in random order - with the best one equally likely to be first, second, third, or last - which of the three methods described would likely be most beneficial to Cher?

You can try solving the problem by assuming the offers are, say, \$600, \$800, \$1000, \$1200. Since there are four offers, you know how many different orders the offers could possibly come in. Out of those, determine how many times method A has Cher accept the best offer, how many times method B is preferable, and how many times method C does the best for her.

# CHAPTER 3 - MATHEMATICS AND THE ARTS

Many people do not realize the visual arts have a strong relationship to mathematics. Consider the two paintings below.

Notice that the first painting appears to be quite flat, where the second painting uses <u>perspective</u>, giving it a more three dimensional appearance. Early Renaissance paintings were the first to make use of perspective. The theory of perspective is mathematically based. Many other principles used in the composition of painting and sculpture are also rooted in mathematics. Some of these are discussed more fully in the following exercises.

## Exercise 1 - Fibonacci Numbers and the Golden Ratio

In the branching of a tree, the following assumptions frequently hold. A branch grows indefinitely but produces no new branch during its first year. At the end of the second year and each succeeding year a new branch is produced by that branch. Starting with a single branch, how many branches are there at the beginning of years 1, 2, 3, 4, 5, and 6? It might be helpful to draw a picture of what the tree looks like each year. Do you find a pattern for the number of branches? (Please answer on next page).

Consider the following sequence of numbers:
1, 1, 2, 3, 5, 8, ... Does this look familiar?

Can you fill in the next 4 terms of the sequence?

Can you describe the pattern?

The sequence you have examined is called the Fibonacci sequence after Leonardo of Pisa, son of Bonaccio (Fibonacci in Italian), a mathematician of the 13th century. An activity at the end of the chapter will help you to learn more about it. It is connected to the visual arts as the following will help you see.

Look at the following rectangles. Which one do you think is the most pleasing aesthetically?

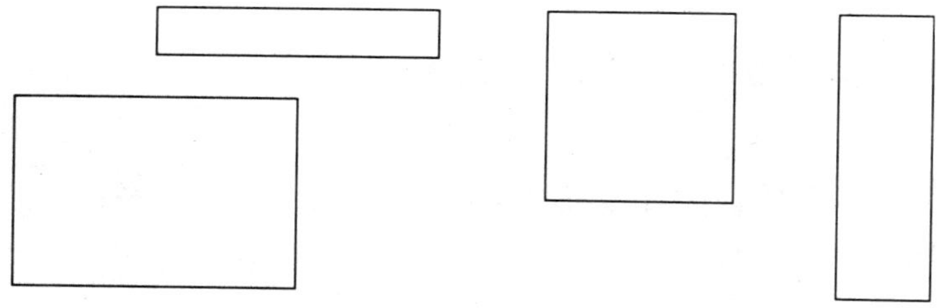

Now consider some familiar objects that are rectangularly shaped. Determine the length to width ratio of a 3 x 5 note card, an 8 x 5 note card, a playing card, the front of a box of cereal, the front of the Parthenon in Athens, Greece (see below), and the figure of St. Jerome in Leonardo da Vinci's painting (also see below).

The number $\frac{1+\sqrt{5}}{2}$ is called the golden ratio and is approximately equal to 1.618. If the length of a rectangle divided by the width equals the golden ratio, the rectangle is called a golden rectangle. Painters, sculptors, and architects recognized long ago the pleasing properties of golden rectangles.

What are some other objects that are golden rectangles?

Note that not all rectangular shaped objects can be golden rectangles. For example, the box that a music CD comes in is square. Why is that?

In many cases, it is not practical to make rectangularly shaped objects in the proportions of a golden rectangle. Find some other objects that are in the shapes of rectangles but not golden rectangles. Can you give some reasons as to why other proportions had to be used?

Recall the Fibonacci sequence: 1, 1, 2, 3, 5, 8, 13, 21,...

Divide the second Fibonacci number by the first.

Then divide the third Fibonacci number by the second.

Now divide the fourth Fibonacci number by the third.

Continue this procedure. What number are you approaching?

## Exercise 2 - Poinsot Stars

Poinsot stars, named for the 19th century French mathematician Louis Poinsot, are interesting designs formed by using specific rules to connect points equally spaced around a circle. Here are some examples:

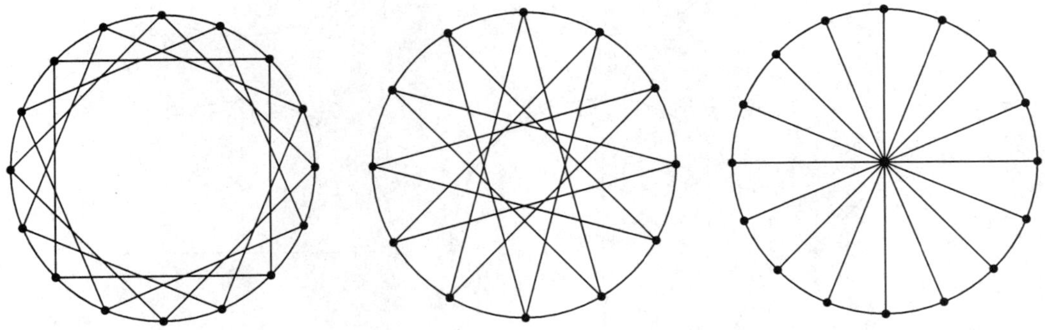

We are going to create some Poinsot stars and look at their properties.

Let's start with a Poinsot star formed from four points. On the next page are lots of circles. Start by placing four equally spaced points on the first circle; put one of the points at the top of the circle. Starting with the top point and proceeding clockwise, connect each point to the next one with a straight line. When you get to the fourth (last) point, connect it to the first (top) point and stop. What you have created is a Poinsot star of order 4; in fact, we will call it a (4, 1) star because i) we used 4 points and ii) from the current point, you just count 1 to find the next point to connect it to. Notice that when you return to the top of the circle you have already touched each point. When this happens, the created star is called a regular star. How to draw the (4, 1) star is shown on the next page.

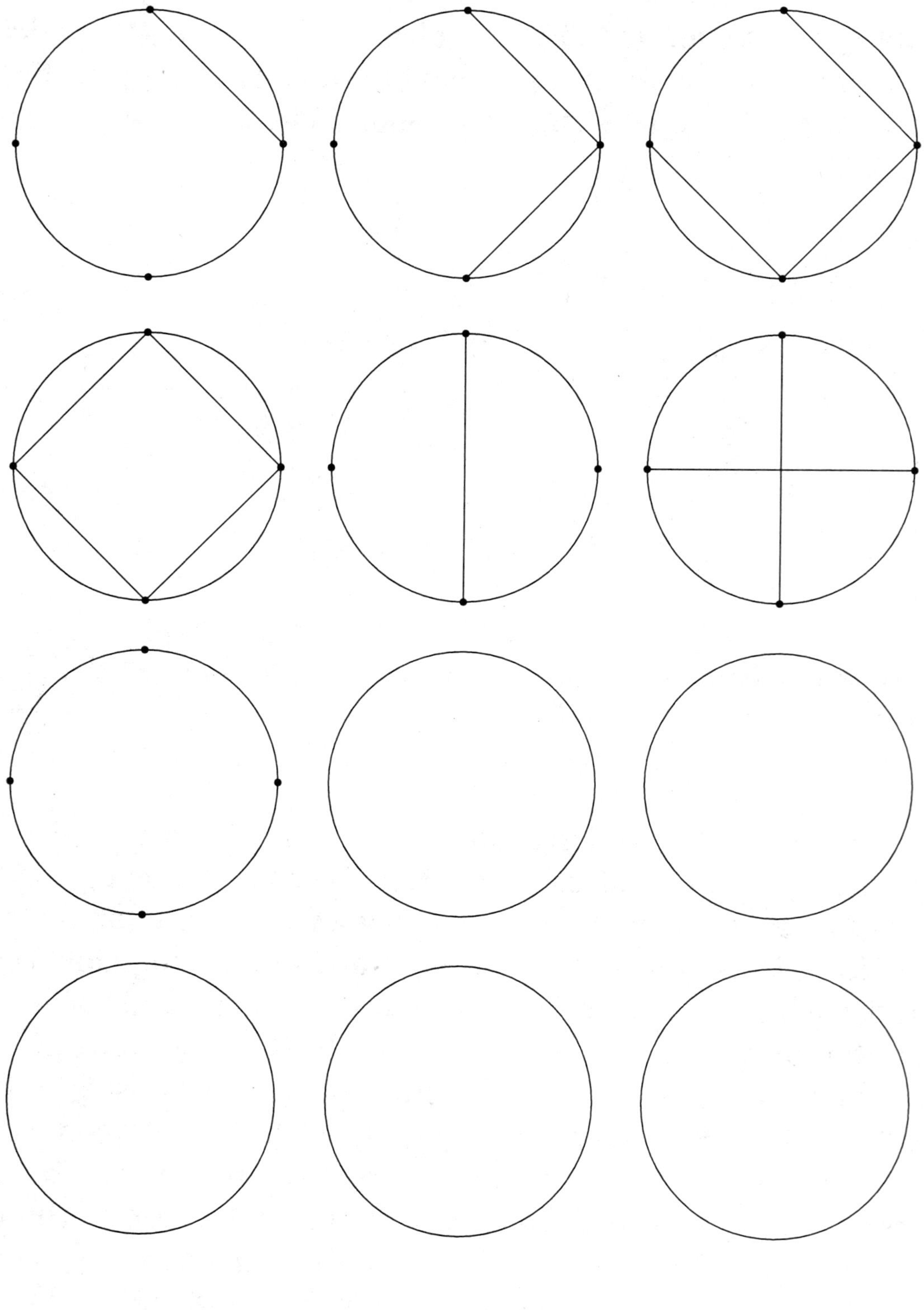

Go to the next circle. Place 4 points on it also. Now, start at the top, proceed clockwise, but count one, two to determine the point whereit should be connected. Make the connection. Count one, two again. Now you are back at the top, and those points are already connected. Notice that you do not touch all the points on the circle before returning to the top. After returning to the top, go to the next unused point and repeat the process. Do this until all points have been used. This is a star which is called a <u>compound</u> star because you returned to the top before touching all points; that is, you cannot draw the entire star unless you lift your pencil off the paper. So the (4, 2) star is compound.

What would the (4, 3) star look like? Try to draw it. Does it look the same as the (4, 1) star?

Now let's consider the stars of order 5. Try drawing the (5, 1), (5, 2), (5, 3), and (5, 4) stars. How many different stars do you get? How many are regular, and how many are compound?

Now find all of the stars of orders 6, 7, 8, and 9. Find the number of different (distinct) stars of each order, and determine how many of them are regular and how many are compound.

Fill in the following chart:

| Order | Total # of Different Stars | # Regular | # Compound |
|---|---|---|---|
| 4 | | | |
| 5 | | | |
| 6 | | | |
| 7 | | | |
| 8 | | | |
| 9 | | | |

Here are some questions that you should consider:

a. How many distinct stars are there of order n?
b. What is the relation between the (n, m) star and the (n, n-m) star? (Hint: Compare the (5, 1) and (5, 4) star and the (5, 2) and (5, 3) star).
c. How do you know if an (n, m) star is compound or regular before you even draw it? (Hint: Look at common divisors of m and n).
d. For some values of n no compound stars of order n exist. What values of n are those?

## Exercise 3 - Modular Arithmetic and Residue Designs

Here are some more interesting designs; these are formed using modular arithmetic. We will learn how to make these residue designs after learning about modular arithmetic.

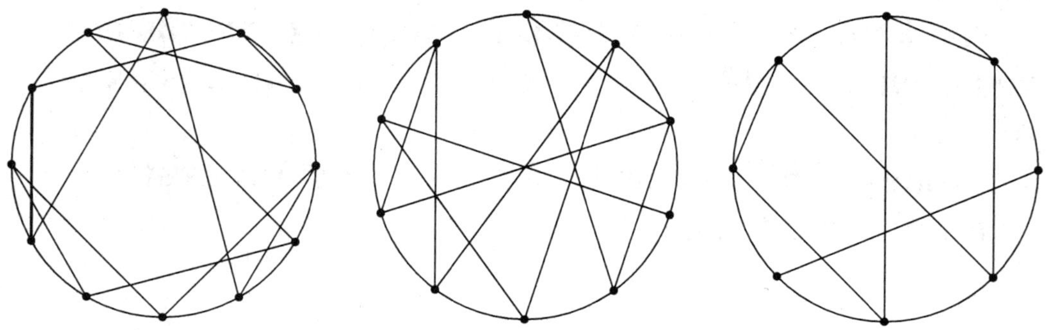

Today is Tuesday. What day of the week will it be in 17days?

In 78 days?

In 129 days? How did you figure this out?

It is now noon. What time will it be in 50 hours?

In 100 hours?

In 275 hours? How did you determine this?

Here is one method you could use in calculating the day of the week it will be so many days from today. You can divide that number of days by 7 and then look at the remainder. If the remainder were 0, then it would be Tuesday, remainder 1 would be Wednesday, and so on up to 6 (what day will it be if the remainder is 6?). Note that you cannot get a remainder larger than 6 when dividing by 7, and that 6 is one less than 7.

The same idea worked for determining what time of day it will be. There we consider the remainder when dividing the number of hours by 24. Our answer is given by considering the remainder, which is an integer between 0 and 23 inclusive. If the remainder is 0, it will be noon, if the remainder is 1 it will be 1:00 PM, etc.

Two positive integers are said to be congruent (modulo n) if each the same remainder is obtained when each number is divided by n. For example, 14 and 29 are both congruent (modulo 5) because both give the same remainder when divided by 5. We write 14 = 29 (mod 5). Notice also that the difference of the two numbers is 15, and 15 is a multiple of 5.

Determine which of the following numbers are congruent (mod 7): (that is, give the same remainder when divided by 7).
3, 13, 14, 18, 22, 6, 15, 34, 42, 28

We can carry this idea out further with our usual operations of addition and multiplication. We are going to consider a new type of addition, which is called "addition (modulo 3)," and is based on regular addition but uses the ideas mentioned above.

The only numbers we use are 0, 1 and 2 (2 is one less than 3) because these are the possible remainders when any number is divided by 3. To perform addition (modulo 3) on two numbers, we first add the numbers as usual, and then divide that sum by 3. Your answer will be the remainder after this division. We use the notation $+_3$ for addition (modulo 3). For example, $1 +_3 2 = 3 \equiv 0$ (mod 3), and $2 +_3 2 = 4 \equiv 1$ (modulo 3). Here is the entire addition table for addition (modulo 3).

| $+_3$ | 0 | 1 | 2 |
|---|---|---|---|
| 0 | 0 | 1 | 2 |
| 1 | 1 | 2 | 0 |
| 2 | 2 | 0 | 1 |

The same thing can be done with multiplication (modulo 3). Here, you take the product of the numbers and then divide the product by 3. Your answer again is the remainder. Here is the table for multiplication (modulo 3) .

| $*_3$ | 0 | 1 | 2 |
|---|---|---|---|
| 0 | 0 | 0 | 0 |
| 1 | 0 | 1 | 2 |
| 2 | 0 | 2 | 1 |

Now try the tables for addition and multiplication modulo 4 and modulo 5.

What is the additive identity for ordinary addition?

Does addition modulo 3 have an identity?

What about addition modulo 4 and modulo 5?

Does every element have an additive inverse?

Is there a multiplicative identity in modular multiplication?

In ordinary multiplication, if the product of two numbers is 0, then at least one of the numbers must be 0. Check your tables to see if this is always true in modular multiplication. In modular multiplication, does every number have a multiplicative inverse?

We will now use modular multiplication to make residue designs. To make an (n, k) residue design: Consider the integers from 1 to n-1. Place n-1 points on the circle, labeling them from 1 to n-1. Multiply each number by k and take the result mod n. Connect each point to its result after multiplication by k (mod n).

Here is an example. We will construct the (7, 3) design.

Place the numbers 1, 2, 3, 4, 5 and 6 around one of the circles on page 31.

Now multiply each number by 3, using multiplication mod 7.

$1 \times 3 = 3 \equiv 3 \pmod 7$
$2 \times 3 = 6 \equiv 6 \pmod 7$
$3 \times 3 = 9 \equiv 2 \pmod 7$
$4 \times 3 = 12 \equiv 5 \pmod 7$
$5 \times 3 = 15 \equiv 1 \pmod 7$
$6 \times 3 = 18 \equiv 4 \pmod 7$

For each line of the above, connect the point labeled with the first number to the point labeled with the last number. You may also want to fill in some of the regions determined by the connections with different colors.

Try the (7, 4) and the (7, 5) designs. Try some others. What ones don't give you much of a design?

## Exercise 4 - Linear Transformations

In mathematics, linear transformations have a variety of uses. These transformations can also be used in other natural and social sciences. In art, they can be used to show how surrealists transform ordinary objects into ones that appear to be distorted.

In high school, you probably studied the x-y coordinate system. You start with a horizontal line (the x-axis) and a vertical line (the y axis) that intersect at the origin. A point's location can be described by an ordered pair of numbers. For example, to find the location of the point (2, 3), go 2 units to the right and 3 units up. After drawing the coordinate axes on the graph paper given at the end of the book, find the point (2, 3). Then find the points (4, 1), (3, 5), (2, 0), and (0, 3). Points may also have negative coordinates, but we will not be using them here.

A <u>transformation</u> is a formula that describes a change made to the coordinates of a point that gives a new point. Let's consider some examples. Fill in the missing coordinates on the right side of tables B and C.

| A | B | C |
|---|---|---|
| $(x, y) \to (x, 2y)$ | $(x, y) \to (3x, y)$ | $(x, y) \to (x+y, y)$ |
| $(2, 3) \to (2, 6)$ | $(2, 3) \to (6, 3)$ | $(2, 3) \to$ |
| $(3, 0) \to (3, 0)$ | $(3, 0) \to$ | $(3, 0) \to$ |
| $(0, 2) \to (0, 4)$ | $(0, 2) \to$ | $(0, 2) \to$ |
| $(4, 1) \to (4, 2)$ | $(4, 1) \to$ | $(4, 1) \to$ |

Now that you have the knack of handling a transformation, let's apply it to some simple pictures.

Draw a rectangle with vertices (0, 0), (3, 0), (0, 7), and (3, 7). Apply the first transformation above to each of these vertices.

Connect the transformed vertices as they were connected before the transformation. What happened to the rectangle?

Try the transformation $(x, y) \to (x, 0.5y)$. What happened?

What happens to a figure under a transformation $(x, y) \to (x, ky)$? ($k$ a constant)

Now apply the second transformation. What happens to the rectangle?

Try also the third transformation. What conclusions can you draw from what you have done?

One reason these are called <u>linear</u> transformations is that straight lines are taken to straight lines under this type of transformation. If the transformation formulas involve such terms as $x^2$ or $xy$, then straight lines may be taken to curves instead of other straight lines. Here's what happens to a line under the transformation $(x, y) \to (x^2, y)$

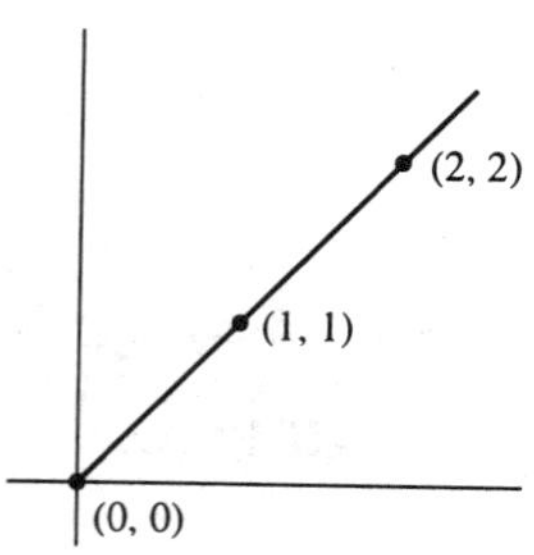


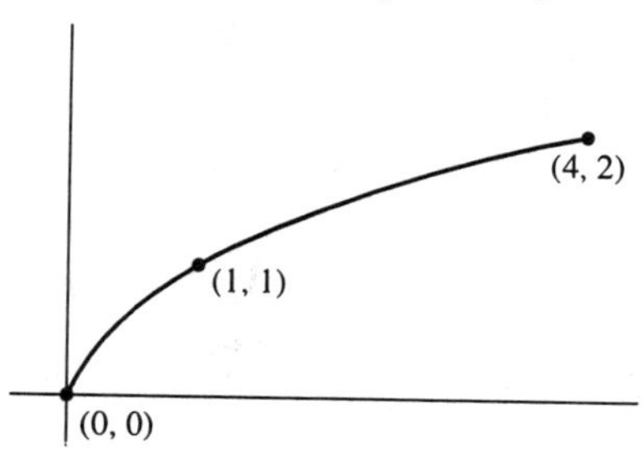

Copy the house below onto a sheet of graph paper. See what happens to it under the above transformations. The coordinates of the house are A(3, 0), B(8, 0), C(11, 0), D(16, 0), E(4, 2), F(6, 2), G(13, 2), H(15, 2), I(8, 3), J(11, 3), K(4, 5), L(6, 5), M(13, 5), N(15, 5), O(5, 10), P(8, 10), Q(11, 10), R(14, 10), S(5, 14), T(8, 14), U(11, 14), V(14, 14), W(3, 16), X(16, 16), Y(9.5, 18).

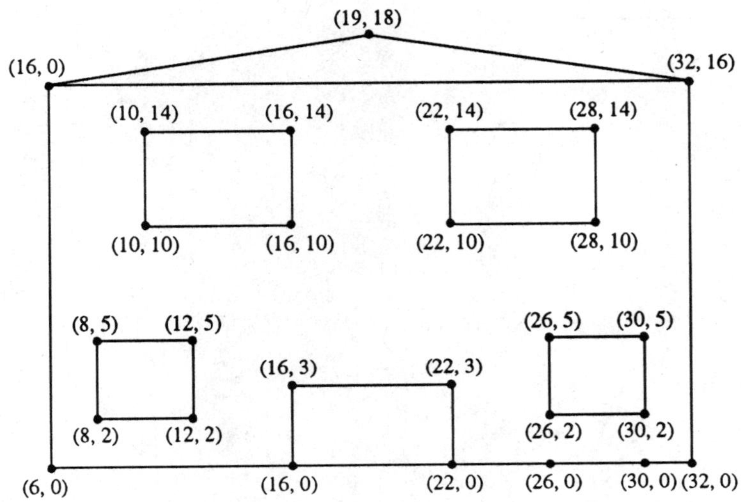


Create some of your own figures. See what happens to them under the above transformations, and then try some of your own transformations.

## Exercise 5 - Tessellations

Look at the following pictures. Do you notice any features common to all of them?

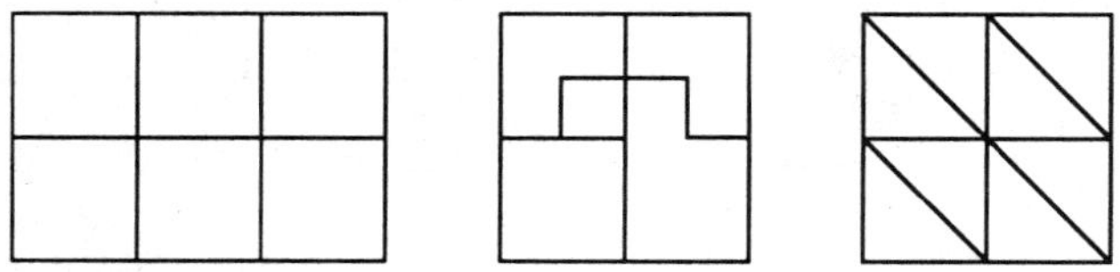

They are all what are called <u>tessellations:</u> figures of exactly the same size and shape that when put together fill an entire region. These can be simple geometric patterns or complicated figures such as done by the Dutch artist M. C. Escher. Tessellations are often used in Islamic art, since this religion prohibits representations of living things in art.

There are many ways to make a tessellation. One of them is described below.

First, draw an equilateral triangle. Draw some kind of curve along one side of the triangle. Copy the same curve along a second side; where the first curve went outside the original triangle, the copy will go inside, and where the first went inside, the copy will go outside. On the third side, mark the middle and draw a curve between the corner of the triangle and this point. Draw the same curve, only reversed, between the other corner of the triangle and this point. Make several copies of the figure drawn below, and show how they fit together.

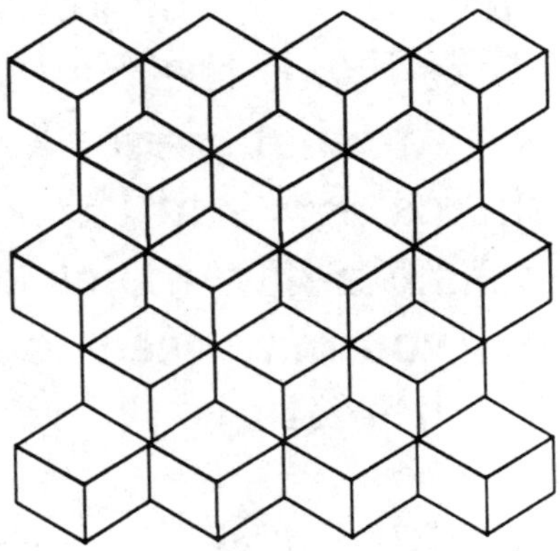

Use this technique, or another, to draw your own tessellations.

## Exercise 6 - Key Signatures and String Vibration Frequencies

Consider the keys on a piano. Here are three octaves.

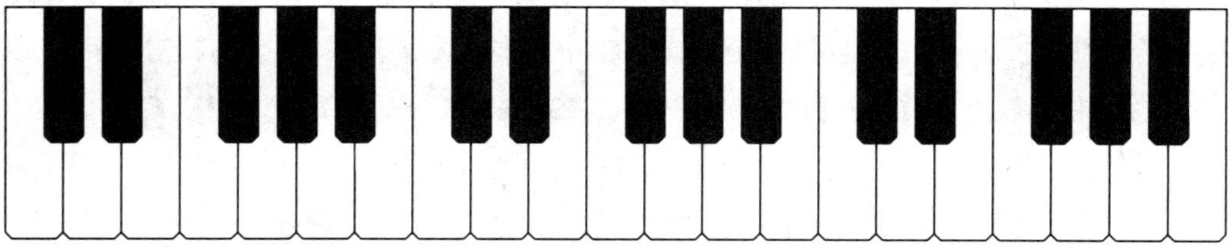

The first note represented is C. There is a pattern that will allow us to determine the sharps in each major key.

The key of C major has no sharps. Count up seven keys, both black and white, from C to G. The key of G major has one sharp, which is F#. Note that F# is the piano key just below G. Count up seven more notes. At what piano key are you now? This major key has two sharps, F# and C#. Continue counting up 7 notes, each time reaching a new major key with one more sharp than the previous key (you may need to use more octaves-note the similarity to modular arithmetic).

The eight major keys with the sharps are:

<u>Key</u> <u>Sharps</u>

1.
2.
3.
4.
5.
6.
7.
8.

The strings on a piano vibrate 440 times per second to produce middle C. To produce the C exactly one octave above middle C, the strings vibrate 880 times per second. At one octave below middle C, the strings vibrate 220 times per second. Can you guess what the frequency of the C two octaves above middle C would be?

Three octaves above?

Two octaves below?

Try to see what you can learn about the key signatures and the flats for the minor keys. Also see what you can learn about the frequencies of the other notes, and how they change according to the octave.

## Exercise 7 - Looking for Structure in Music and Literature

Like mathematics, most literature and music follows a logical structure. No doubt you have studied forms of poetry such as sonnets, haiku, and limericks. These all follow very specific rules of rhyme and/or meter. Music is often organized in a similar fashion.

Listen to the song "Yesterday" by the Beatles. What is the rhyme scheme in the first verse?

Now listen to the second verse. Is the rhyme scheme the same as the first verse? What about the music itself? Is it the same as the first verse?

Now listen to the next portion of the song. Does the rhyme scheme change? How about the music?

Listen to the rest of the song. Note for each part if it repeats a previous part in rhyme scheme and/or music. Describe the structure of the song.

Try the same thing with some of your favorite songs. Here's a diverse collection; you might consider some of these.

Rush- "Freewill"
The "Phantom of the Opera" Original Cast- "Music of the Night" (Note: The lyrics are different on the highlights version).
Ace of Base - "The Sign"
Queen- "Crazy Little Thing Called Love"
Andy Williams- "Moon River"
Buddy Holly- "Peggy Sue"
Linda Ronstadt and Aaron Neville- "Don't Know Much"
OMD- "If You Leave"
Metallica- "Enter Sandman"

Classical music is also highly structured. Why not try an introduction to music class?

## Activity 1 - The Fibonacci Numbers and the Golden Ratio

Use a piece of graph paper to construct a golden rectangle. Here's how:

1. Turn the paper so it is longer horizontally than vertically. Draw a square 20 blocks on a side as close to the left side of the paper as you can.

2. Place the point of a compass in the middle of the bottom of the square. Put the pencil on the upper right hand vertex of the square. Draw an arc from that point down to the same line that the bottom of the square is on.

3. Extend the bottom of the square to the point where the arc intersected that line. Use that as the base of the golden rectangle. The height of the rectangle is the same as that of the original square. Divide the length of the rectangle by its height. Are there any other golden rectangles in your picture?

## Activity 2 - You the Artist

In the manner discussed in these exercises, create your own designs using the techniques of coordinate transformation, tessellation, Poinsot stars and residue designs.

## Activity 3 - Surrealism

One of the most famous surrealists was the artist Salvadore Dali. Find a reproduction of his painting, "The Persistence of Memory." How does this painting relate to the Exercise 4 - Linear Transformations? Write a brief paper on the life and works of Dali.

Rene Magritte was another very well known surrealist painter. Read about his life and works, and write a brief paper.

## Activity 4 - Golden Rectangles

Write a short paper on how golden rectangles have been used in art and architecture. Some artists whose work you might consider include Seurat, Gris, and Le Corbusier.

## Problems to Try

1. Draw a circle and construct all Poinsot stars of order n on the same circle. The circle is divided into regions separated by the lines in the stars. How many distinct regions are there?

2. Mr. P. Leonardo breeds a strain of self-fertilizing worms that reproduce in the following manner. Each worm produces its first offspring at the age of two weeks and continues to produce exactly one offspring each week thereafter. If Mr. Leonardo started breeding these worms just over ten weeks ago, and if he had exactly two newborn worms at that time, how many worms does Mr. L have now, assuming that none of them have died?

3. What would happen if a pair of rabbits were put into a walled enclosure to breed?
The situation can be discussed in the following way: How many pairs of rabbits will there be after a year if it is assumed that every month each pair produces one new pair which begins to bear young two months after its own birth?

4. A male bee develops from an unfertilized egg; in other words, it has a mother but no father. A female bee develops from a fertilized egg; it has both a mother and a father. Consider the family tree of a single bee. Find the number of direct ancestors of a female bee and of a male bee.

5. Let $f_1, f_2, f_3, \ldots$ be the Fibonacci sequence.
Compute $(f_4)^2 - (f_3)^2$ and $f_2f_5$
$(f_5)^2 - (f_4)^2$ and $f_3f_6$
$(f_6)^2 - (f_5)^2$ and $f_4f_7$
Try some others. Make a conjecture concerning this. Prove your conjecture.

Now compare $(f_4)^2 - (f_3)^2$ and $f_3f_4$
$(f_5)^2 - (f_4)^2$ and $f_4f_5$
$(f_6)^2 - (f_5)^2$ and $f_5f_6$

Make another conjecture.

6. Describe the pattern and continue the sequence:
8, 9, 7, 10, 6, 11, 5, 12, .... .

7. Is 144 the largest perfect square Fibonacci number?

8. Albrecht Durer has a magic square in his woodcut, "Melancholia." Find a reproduction of the woodcut. What is a magic square?

# CHAPTER FOUR - COUNTING ARGUMENTS AND PROBABILITY

What does the probability of something happening mean to you? You often hear remarks like the following on the radio or TV, or read them in the newspaper: "The chance of rain tomorrow is about 30%." "The odds that the Chicago Cubs will win the World Series this year are about 1000 to 1." "The probability of winning the lotto is about 1 in 4 million." Do you understand what is being said in these sentences? Do they make sense? The following exercises will help you understand more about probability. To understand basic probability arguments, however, we must first discuss methods of counting the number of possible results for specific events.

## Exercise 1 - The Multiplication Principle

Suppose you are traveling from Pittsburgh, PA to Washington D.C. You have a choice of three possible routes (numbers 1, 2, and 3) from Pittsburgh to Breezewood, and two (label them A, B) from Breezewood to Washington. How many routes are there? You can figure this out by writing them all down. One route would be 1, A.

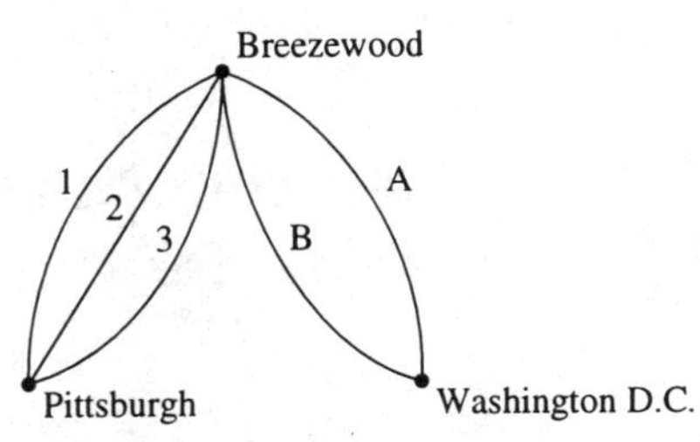


(You should have an answer of 6).

How would your answer change if there were four possible routes from Pittsburgh to Breezewood?

What if there were three routes from Pittsburgh to Breezewood and three from Breezewood to Washington, D.C.? Can you see how a problem like this would be solved if there were 30 routes and 18 routes, respectively?

If there are m ways to do Step 1 of a problem and n ways to do Step 2, how many possible ways are there to do the entire two step problem? Use your work above to figure this out.

Let's try another problem. Flip a coin once. How many possible outcomes are there?

Flip a coin twice. How many sequences of heads and tails are possible for the two flips?

What if the coin is flipped three times?

n times? How did you figure this out?

The above leads to a very important principle that we can use when trying to count how many ways a particular task can be done. It is called <u>The</u> <u>Multiplication</u> <u>Principle</u>.

"If a task is to be completed by doing several consecutive steps, then the total number of possible ways to do the entire task is to multiply together the total number of ways to do each step." Use what you have learned to answer the following questions.

1. "Dear Abby" and "Ann Landers" appear side by side in many daily newspapers. Each one wants to run only one long letter on April 1. Suppose Abby's column is always on the left. If Abby has 5 letters to choose from and Ann has 4, in how many different ways could the columns appear?

2. You roll a red die and a green die. How many different possible results are there?

3. You are given a true/false test with four questions. Since you have no idea what any of the answers are, you guess the answer to each question. In how many different ways could you fill out the answer sheet?

What if there were five questions?

n questions?

Answer the same questions except now assume the test is multiple choice and there are 4 possible answers for each question.

4. You are going to create your own sundae at Basler's. You have a choice of one of 31 flavors for the ice cream, a choice of one of 5 toppings, whether to have nuts or not, and whether to have whipped cream, jimmies, or neither of these. How many possible sundaes could you create?

5. Pan's Restaurant once claimed there were 256 different ways to order a single patty hamburger at their restaurant. How could this have been possible? Hint: Consider different possible condiments for the hamburger, and see that either each condiment could be on or not on the hamburger.

## Exercise 2 - Permutations

Let's move on to putting a set of objects in order. These questions can still be answered using The Multiplication Principle.

1. In Major League Baseball, there are five teams in the Central Division of the National League. Assuming no ties, how many different possible rankings are there for the teams in the division at the end of the season? (Hint: How many teams could have ended up in first place? Once first place has been filled, how many teams remain that could have ended up in second place?)

2. A cake recipe with six ingredients will only turn out properly if the ingredients are added in a certain order. If the part of the recipe that explains the order to be used has been lost, in how many different possible orders could the ingredients be added?

This leads us to the question of how many ways there are to order a set of n items. First, let us define n factorial, which is denoted by n!.

3! (read "3 factorial") =3 x 2 x 1 = 6
4! = 4 x 3 x 2 x 1 = ______
5! = ________________= ______
6! = ____________________ = ________
n! = ______________________________

From this definition and problems 1 and 2 above, can you determine how many ways there are to order a set of 5 objects?

6 objects?

n objects?

Now answer the following questions.

3. Nine horses are in the Kentucky Derby. How many possible orders are there for the horses to finish?

4. Four candidates are running on the Democratic ticket in the mayoral primary. Assuming no ties, what is the number of possible orders in which the candidates can finish?

n! is also denoted by $P_{n,n}$ which means the number of ways to choose n objects from n and then put them in order.

5. Return to problem 7 above concerning the horse race. In how many different ways can the horses finish in the money (be in the top three)?

6. The DJ at a classic rock station wants to play six Led Zeppelin songs in a row. He wants to choose these from his ten favorites. In how may ways can he fill out the time it will take to play these songs?

The answer to Problem #5 can be written $P_{9,3}$ and the answer to Problem #6 can be written $P_{10,6}$. What do you think this notation means?

$P_{n,k}$ is read as the number of permutations (meaning arrangements) of n objects taken k at a time. How is $P_{n,k}$ calculated? State your method in words, and then try to express the answer as a formula.

Calculate $P_{5,3}$ $P_{6,4}$, and $P_{7,3}$.

Try answering the following questions. Give a numerical answer and the answer in the P notation (if appropriate).

6. In how many possible ways can 6 people be put in line for a photograph?

7. In the first U.S. presidential elections, the person receiving the most electoral votes became president and the person receiving the next largest number of votes became vice-president. If there were five candidates in one of these elections, in how many ways could the offices of president and vice-president have been filled?

8. How many "words" of 6 letters can be made from the letters a, b, c, v, e, and x if each letter must be used once? How many 6 letter words can be formed if letters can be used more than once in the same word?

b) How many 3 letter words can be made from the above letters if no letter may be used more than once?

## Exercise 3 - Combinations

Let's consider a similar problem, but where the order of the objects chosen does not make any difference. Here is an example.

Suppose there are five people: Ashley, Bette, Carla, Daniel, and Ed. Suppose you are choosing 2 of these 5 people to form a committee. Then, the order the people are chosen doesn't make any difference. How many different committees could you choose?

You have chosen 2 objects from 5, where the order in which the objects were chosen does not make a difference. This denoted by $C_{5,2}$, which is the number of combinations of 5 objects taken 2 at a time. How are $C_{5,2}$ and related (recall that $P_{5,2} = 20$)?

Let's now consider a game of cards where you are dealt 3 cards. If the order in which you received the cards were important, the number of possible hands would be $P_{52,3}$. But if you are allowed to pick up the cards, look at them, and rearrange them, the order in which they were chosen does not make a difference. So, for example, receiving QH, 9D, 3C is the same as 3C, QH, 9D or any other order of the same 3 cards. So the number of possible hands is $P_{52,3}/3!$ , which we denote by $C_{52,3}$.

From these two examples, can you determine how $C_{n,k}$ and $P_{n,k}$ are related? $C_{n,k} = P_{n,k}/?$ ? (Hint: 2 = 2!)

Calculate $C_{5,3}$, $C_{6,4}$, and $C_{7,3}$.

Try answering the following questions.

1 a) A history major at Agony College must take any 11 courses of the 20 offered by the department. If Julianne is a history major, in how many ways could she select the courses to complete her major?

b) Suppose 9 of the courses are numbered below 300. Julianne must take 5 courses numbered below 300 and 6 numbered above 300. Now in how many ways can she complete her major?

2. A poker hand contains 5 cards. Find
a) the number of possible poker hands.

b) the number of hands containing only black cards.

c) the number of hands with no aces.

d) the number of hands containing exactly one king.

## Exercise 4 - Probability and Multiple Repetitions of an Experiment

You "probably" know that if a coin is flipped, the likelihood of getting a head is the same as the chance of getting a tail. What happens if the coin is flipped a large number of times? Will you always get half the flips to be heads and half tails? If you flip the coin 100 times, and then flip it again 100 times, will you get the same results?
What do you think?

Let's try the following simple experiment. Flip a coin ten times; note how many heads and how many tails you get.

Is this what you expected? Are you surprised at the results?

Do the experiment again (to do it more quickly, you can take 10 coins and just drop them on the floor). Now how many heads and how many tails did you get?

Do the experiment 15 times altogether. For each trial, indicate how many heads you obtained and how many tails you obtained.

Now compile results for the entire class in the following way:

1. Determine the total number of coin flips that were carried out. How many heads and how many tails were obtained?

2. Divide the total number of heads by the total number of flips, and the total number of tails by the total number of flips. Did each of these come out to be close to $1/2 = 0.5$?

3. Now determine the total number of sets of 10 flips which were performed. Make a chart as follows:

| Number of Heads in the 10 Flips | Number of Trials |
|---|---|
| 0 | |
| 1 | |
| 2 | |
| 3 | |
| 4 | |
| 5 | |
| 6 | |
| 7 | |
| 8 | |
| 9 | |
| 10 | |

If, when the coin was flipped ten times, 6 heads were obtained 30 times, then put 30 in the second column in the same row as the 6. Make sure the second column adds up to the total number of trials that were performed.

4. Use graph paper to construct a histogram showing the class results. What is the shape of the histogram? Below is a histogram obtained when this experiment was carried out in one class.

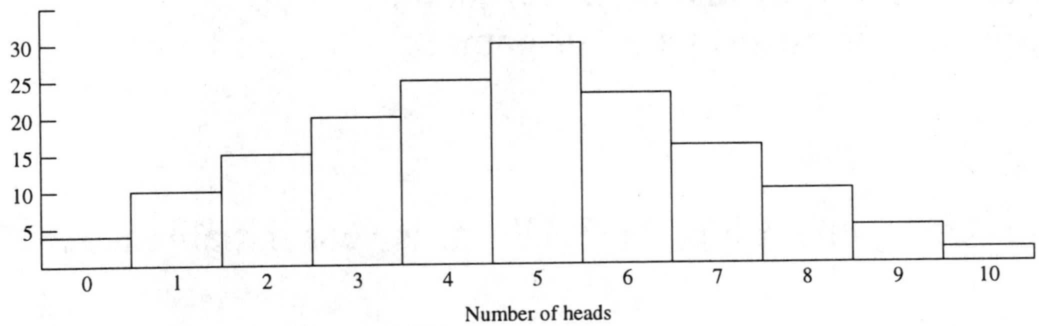


Estimate the probability of obtaining 0 heads in 10 flips, 1 head in 10 flips, etc., by dividing the numbers in the second column of your chart above by the total number of times the experiment was performed. Save these results and compare them to the theoretical ones, which will be discussed in class.

What percentage of the time did you get exactly 5 heads and 5 tails? Is that what you would have expected before you did this experiment? Is 5 of each more likely or 4 of one and 6 of the other?

## Exercise 5 - Equally Likely Outcomes

You and your friends Joe and Joanne are feeling lazy. You decide to play the following game. Each of you puts in 1 cent. Two nickels are tossed. You tell your friends: "If 2 heads come up, Joe gets the three pennies. If 2 tails come up, Joanne gets the three pennies. If one head and one tail come up, I win the 3 pennies. Since these are the three possible outcomes, this should be a fair game."
What do you think is meant by a fair game?

Do you think this game will be fair? What is your initial reaction?

Divide into groups of three people. Have one person be Joe, one Joanne, and one the speaker. Have each group play the game 20 times. What happened in your group? Has one person won more pennies than the others?

Compile class results: determine how many pennies were won by Joe, how many by Joanne, and how many by the speaker.

Do you think this game really is fair? If not, approximately what is the probability that Joe will win the game (use the class results from above and the total number of pennies won.)?

Joanne?

The speaker?

Find the <u>odds</u> of each person winning by considering the ratio of the number of times that person won to the number of times that person did not win.

Can you explain why this game is not fair?

How could the rules of the game be changed so that the game is fair?

## Exercise 6 - Equal Probabilities?

A "friend" makes the following challenge. "I have here three cards. One is black on both sides, one is white on both sides, and one is black on one side and white on the other side. I am putting the cards in this envelope. You draw one, but do not look at the other side. If the other side of the card is the same color, you give me $1. If the other side of the card is a different color I give you $1. This is a fair game." Do you think this game is fair? Write down your initial reaction.

Divide the class into pairs. Make cardboard cards fitting the conditions of this problem. Play the game 15 times for each pair. What happened in your pair? Do you think it is a fair game?

Now compile class results. Do you think this is a fair game? Why or why not?

## Exercise 7 - Independent Events

Many roulette players believe that if a roulette wheel comes up red twice in a row that the next spin is much more likely to yield black than red. Since we do not have a roulette wheel in class, let's model a similar but simpler problem. If a die is tossed and an even number shows twice in a row, is it then more likely that an odd number will come up on the third roll of the die? Let's perform an experiment to test this theory.

Roll a die twice. If you get two odd numbers in a row, roll the die again and note if the third number is odd or even. If you get two even numbers, roll the die again and note if the number is odd or even. If the first two rolls are neither both odd nor both even, then don't count this trial. Perform the experiment until you have 20 trials that count.

How many times were two even numbers followed by an odd number?

By an even number?

Is there much of a difference?

How may times were two odd numbers followed by an even number?

By an odd number?

Again, was there much of a difference?

Answer the above questions for the whole class, and comment on the results.

What did the experiment have to do with the roulette wheel example mentioned at the beginning of this exercise?

## Exercise 8 - Expected Value

Try the following game. You flip two coins. Your wealthy friend agrees to pay you \$2 if 2 heads come up, \$1 if 1 head comes up, and nothing if 0 heads come up.

Divide the class in pairs and play the game 15 times for each pair. How much money did you win?

On average, how much money would you expect to win each time you play the game?

Compile the results for your whole class.

Now do the following calculations. Let P(0)be the probability of getting 0 heads, P(1) be the probability of getting one head, etc.

P(0) x \$0 = (____) x 0 = (____)
P(1) x \$1 = (____) x 1 = (____)
P(2) x \$2 = (____) x 2 = (____)

Add these products and obtain ________

How does this sum compare to what you calculated previously?

What you have just calculated is the expected value of this game. Each possible amount won is multiplied by the probability of winning that amount. These numbers are all added to give the expected value, which is the amount you expect to win, on average, if the game is played many times.

Try the following to see if you understand the principle of expected value.

Roll a die. If a 1 is rolled, you win \$1, if a 2 is rolled, you win \$2, etc. What is the expected value of your winnings if you play this game? If you have to pay \$3 to play, would you?

Play the same game, only use two dice instead of one. Now what is the expected value of your winnings?

## Exercise 9 - A Dicey Game

A game called "Dicey" is played with a set of nonstandard dice whose faces are shown flattened out. The game is played by two players. Each player chooses a different die and the players each roll their die at the same time. The player with the greater number showing on his or her die wins the game. If you were to play "Dicey," would you want to choose your die before your opponent or after your opponent has chosen? Why? Let's perform an experiment to help us answer this question.

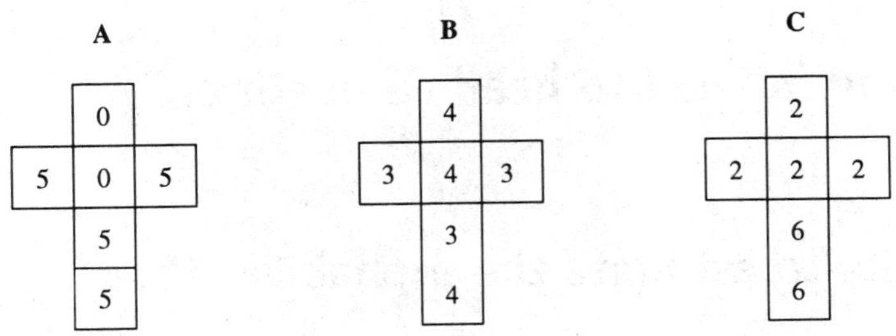


Make some cardboard dice to represent each of the dice above. Divide the class in pairs. Have each pair play 10 rolls of die A against die B, 10 rolls of die B against die C, and 10 rolls of die C against die A. What happened in each case?

Compile results for the entire class.

Which die will win more often in head to head competition?

Use your experimental results to estimate the probability that
i) die A beats die B

ii) die A beats die C

iii) die B beats die C

When you have learned how, calculate the theoretical probabilities for the above.

## Exercise 10 - The World Series

The following problem will model the World Series (You don't have to know anything about baseball-all you need to know is that the American League champion plays the National League champion, and the first team to win four games wins the Series). Assuming that the two teams are evenly matched (and that there is no home-field advantage), what is the probability that the series will be over in 4 games? 5 games? 6 games? Or go all 7 games? We are going to carry out an experiment in an attempt to answer these questions.

We will model the problem in the following way. Divide the class into pairs. Each pair will perform the experiment. One member of each pair will flip a coin. If heads comes up, the National League team wins the first game. If tails comes up, the American League team wins the first game. Flip the coin a second time to determine which team wins the second game. Continue to flip the coin until one of the teams has won four games. Note which team won the series and how many games the series took. Perform the experiment 25 times. Compile results for your experiment. This type of experiment is called a simulation.

Now compile results for the entire class.

Find the experimental probability that the National League team will win the series. How does that compare to the theoretical probability?

Find the experimental probabilities that the series will go 4 games, 5 games, 6 games, and 7 games from the class results. Compare these to the theoretical probabilities.

In how many of the series did a team come back from a 3-1 deficit and go on to win the series?

From 1905 to 1974, one team had a 3-1 lead 33 times, and only three times did the team behind go on to win. How does that compare with your experiment?

Did any team with a 3-0 deficit ever go on to win the series?

We can also model the series if the teams are not evenly matched. Suppose the NL team is twice as likely to win a particular game as the AL team. Instead of flipping a coin, roll a die. If a 1, 2, 3, or 4 comes up, then the NL team wins that game. If a 5 or 6 comes up, then the AL team wins. Roll the die until one of the teams has won 4 games. Again, note which team won and how many games the series took. Perform the experiment 25 times per pair. Compare your results to the theoretical results.

Explain why the models used are appropriate for the problem.

## Exercise 11 - The Birthday Problem

The probability that at least two people in a group will have the same birthday increases with the number of people in the group. It is higher than you might think. Guess how many people you would have to have in a group to make the probability at 50% or more that at least two people share a birthday (month and date only).

Then consider this exercise. The Multiplication Principle will be useful here.

1. a) Find the number of ways two people can have a birthday whether they are the same or not. (Ignore leap years).

   b) What is the number of ways two people can have different birthdays?

   c) What is the probability that the two people do not have the same birthday? To find, divide the answer in b by the answer in a).

   d) What is the probability that the two people have the same birthday?

2. a) Find the number of ways three people can have a birthday whether they are the same or not.

b) What is the number of ways three people can have different birthdays?

c) What is the probability that no two people among those three would have the same birthday?

d) What is the probability that at least two people among these three share a birthday?

3. a) Find the number of ways n people can have a birthday whether they are the same or not.

b) What is the number of ways n people can have different birthdays?

c) What is the probability that no two people among those n would have the same birthday?

d) What is the probability that at least two people among these n share a birthday?

e) For what value(s) of n are the answers to c) and d) close to 50%?

Using the same ideas as above, find the probability that at least two people in a group were born on the same day of the week, if the group consists of a) 3 b) 4 c) 5 d) 6 people.

## Exercise 12 - How Likely is a Match?

Hilary decides to give a housewarming party. She writes personal invitations and addresses all the envelopes at her desk by an open window. When she leaves the room a strong wind blows the invitations and envelopes all over the room. Later, her secretary comes to stuff the envelopes. He doesn't realize the invitations are individualized and so he simply stuffs invitations randomly into the envelopes. If there are thirteen invitations sent, what is the probability that none are correctly matched? Make a guess.

Now estimate this probability by performing an experiment which models the given situation. Use two full suits of a standard deck of 52 playing cards. Use one suit to represent the invitations, and another to represent the envelopes. Shuffle each suit thoroughly but separately. Then choose one card from each pile. Do they match? If not, choose another card from each pile. Keep doing this until either you get a match or you have used all of the cards. Perform the experiment 20 times per pair of students, noting whether there was a match or not. Compile class results to determine the experimental probability of obtaining at least one match.

It is not too difficult to determine the theoretical probability of getting at least one match if there are 2, 3, 4, or even 5 invitations and envelopes. Try to do this. If you have access to a computer, you can try the problem with even a larger number of invitations and envelopes.

## Exercise 13 - Lotteries

Many states run various types of lotteries. We will look at the probabilities of winning in different types of lotteries.

1. One very common type is the "Daily Number." A three digit number is chosen for the day. To win, you must guess what number will be chosen.

a) If any of the three digits can be 0-9, how many different three digit numbers are there that could be drawn?

b) What is the probability of winning on a specific day?

c) If you have to pay $1 to play, and win $500 if your number is chosen, what is the expected value of the lottery for you?

d) Some states will pay you $50 if you "box" the number you have chosen. That means any permutation of the three digits is also a winner. What is the probability you will win if you "box" your choice?

What is the expected value of this option?

2. A similar type of lottery involves choosing a 4 digit number. Answer questions a) and b) above for this game.

3. In another type of lottery, six numbers from among the numbers 1-40 are chosen. You win only if you correctly predict the six numbers. How many ways are there to choose six numbers from 40?

What is the probability that you will win this lottery?

A small prize is often given if you correctly pick five of the numbers. What is the probability that you will win one of these prizes?

What if you need to predict only four numbers correctly to win?

4. In another type of lottery, eleven numbers from among the numbers 1-40 are chosen. You choose seven numbers. If the numbers you choose are among the seven chosen, you win. What is the probability that you will win this lottery?

Here are some additional problems to try.

1. The Joneses have been married for one year. They plan to start a family, and really want a girl. They plan to have children until a girl is born or until they have 4 children. What is the probability that their wish will come true? (Assume boys and girls are equally likely).
The Smiths want a boy, but they will stop after 3 children. What is the probability that their wish will come true?

2. "Bedtime Crumblies" is testing a recipe for a new line of snack cracker. 30 people each sample four recipes labeled A, B, C, and D and rank them 1st, 2nd, 3rd, or 4th in order of taste preference. Recipe D received no 1st place votes; recipe A received no 3rd place votes; all preferred B to C; 1/6 preferred D to A; 1/2 preferred B to A; 1/5 preferred D to B; only 2 put D in last place.

   a. List all possible rankings of the 30 tasters.
   b. How many tasters recorded each of the possible rankings if no two rankings received the same number of votes?

3. You and a friend remove your hats, and a third person places in each of them seven playing cards, ace through 7 (suits are irrelevant). You and your friend turn your backs while the third person mixes the cards in each hat and then randomly picks five cards from your friend's hat and three cards from yours.

Now you must each guess out loud which cards were removed from your own hat (irrespective of the order in which they were picked). The whole process will be repeated until one of you guesses all the removed cards correctly. Although your friend has to guess five cards while you have to guess only three, he nevertheless offers you even money that he will correctly guess all the cards removed from his hat before you guess all the cards removed from yours. Is it a fair bet?

4. It is a well-known fact that if you flip a coin enough times it will, on average, land heads up half the time. The more times it's flipped, the closer the probability comes to exactly 50 percent. Suppose somebody offers to bet you even money that in at least one of three sets of 20 flips each, the results will be exactly 10 heads and 10 tails. Are you game?

5. It has been claimed that the 13th of the month is more likely to fall on a Friday than any other day of the week. Is this really true? Discuss this.

6. Determine which is more likely; that 20 heads will be obtained if a coin is flipped 20 times or if a ticket purchased in Lottery 3, Exercise 11, will be a winning ticket.

## Activity 1

Find another type of lottery that is held by your state or a fundraising organization. Find the probability of winning the grand prize and any other prizes awarded. Also find the expected value of the game.

## Activity 2

Find the probability of winning the various prizes in a sweepstakes such as Reader's Digest or Publisher's Clearinghouse. You'll have to do research to determine how many entrants there are.

# CHAPTER FIVE - STATISTICS

You have often heard statements like the following: "Two out of three doctors recommend... " "The President's approval rating rose four points to 45%." "The average per capita income for a resident of the United States is $22,378." "If the election were held today, Sam Smith would win with 60% of the votes cast." "Jay Bell's average rose from .221 to .273 during the month of July, when he hit .421."

The word statistics has at least two related, but different, meanings. Statistics are numbers which measure quantities, e.g. a person's height, a baseball player's batting average, or the percentage of registered voters who cast a ballot in an election.

Statistics also refers to the methods used to gather and present numbers such as those listed above.

### Exercise 1 - Inferential vs. Descriptive Statistics

The two meanings of statistics mentioned above often come together when we read about a President's approval rating, the national unemployment rate, the Nielsen ratings for television programs or the number of college students who do not know why we have summer and winter. When we read, see, or hear statements which include such statistics we must carefully scrutinize the information we are given to determine whether we are dealing with descriptive statistics or inferential statistics.

Descriptive statistics is concerned with collecting and describing a set of numbers so as to yield meaningful information about that set of numbers. No inferences are made about a larger set of numbers or broader context.

Examples: a baseball player's batting average
your quality point average

Can you give some more examples?

<u>Inferential</u> <u>statistics</u> is concerned with collecting and describing a set of numbers in order to infer information about a larger set of numbers in a broader context.

Examples: the President's approval rating
pre-election political polls

Can you think of some other examples?

Try this mini-quiz. For each of the following statistics, determine if it is more likely to be an example of inferential or of descriptive statistics.

_____a) the number of people in the U.S. who prefer Coke to Pepsi.
_____b) the number of sociology majors at your college.
_____c) the number of credits you have earned at your college.
_____d) the number of college students who consider themselves liberals.
_____e) the number of people who exercise more than twice per week.

Here is an exercise for you to complete outside of class and bring to your next class meeting.

Find an article in a newspaper or magazine which reports inferential statistical data. Identify phrases relating to the reliability of the inference. Remark about whether or not sufficient information is given to justify the inference (this is your own opinion).

In January 1990 television station WTAE 4 (Pittsburgh) reported on their evening news that prices for gasoline and diesel fuel were increasing and that one consequence of this was an increased operational cost to Pittsburgh's public transit system (PAT). The report stated that each 1 cent per gallon increase in the cost of fuel caused a $100,000 per year increase in operating costs to PAT. What additional information would you need in order to determine the significance of these figures?

## Exercise 2 - Mean, Median, and Mode

It is often useful to describe a set of numbers by one single number which represents the "average" value, or the "center." The three measures of the center that are most often used are the mean, the median, and the mode.

Many people think of the <u>mean</u> when they hear the word "average." The mean of a set of numbers is found by adding up all of the numbers in the set and then dividing by how many numbers are in the set.

Use this definition to find the mean of the following test scores: 88, 93, 78, 91.

The <u>median</u> of a set of data is found by arranging the numbers in order of magnitude and finding the middle value.

If there are an odd number of values in the set, it is easy to find the median.

Find the median of the following bowling scores by first arranging them in increasing order and then finding the middle value: 145, 220, 175, 89, 164, 122, 212, 210, 115.

If there are an even number of values then there are two values in the middle; add these two middle values and divide by 2 to obtain the median in that case.

Find the median of the following ages of Jean's children:
38, 15, 23, 28, 34, 19, 30, 27.

The mode is the value which occurs most often in the set of data. If no value occurs more than once, then there is no mode. It is also possible for there to be more than one mode. A data set with two modes is called bimodal.

Find the mode (or modes), if any, of each of the following:
a) {4, 23, 3, 23, 4, 6, 18, 20, 28, 4, 8, 23, 4}

b) {4, 9, 15, 8, 17, 12, 14, 22, 10}

c) {3, 18, 15, 10, 15, 12, 10, 9, 10, 15}

Each of these numbers give some useful information about the set of numbers it describes. The mean, median, and mode are all examples of descriptive statistics.

Let's talk more about these measures of "center." A randomly chosen group of first year health professionals (not doctors) who work in Cleveland were asked to list their salaries.
Here are their responses:

| | | | | |
|---|---|---|---|---|
| $42,000 | $29,000 | $23,000 | $26,000 | $27,000 |
| $24,000 | $27,000 | $30,000 | $52,000 | $38,000 |
| $24,000 | $27,000 | $28,000 | $30,000 | $45,000 |

Find the mean, median, and mode of this set of numbers and discuss which best describes the values. Give the good points and drawbacks of each. If you were to look for a job in this field, what would you expect your first year salary to be and why?

## Exercise 3 - Spread of Data

The "center" or "average" of a set of numbers is a very useful statistic, but it by no means completely describes the set of data.

Consider the following sets of data:

Set 1 {1, 24, 50}     Set 2 {22, 24, 29}

Find the mean and median of each set of data (there is no mode for either). How do they compare?

Describe an obvious difference in the sets of data.

One way of measuring that difference is by finding the range of the data set. To find the range, subtract the smallest element from the largest. Do this for the above sets.

Let's try this problem.

There is a certain neighborhood in Santa Teresa, CA, where the annual household incomes of the residents are listed below: (Call this Data Set A)

| | | | | |
|---|---|---|---|---|
| $45,000 | $54,000 | $58,000 | $45,000 | $120,000 |
| $60,000 | $62,000 | $53,000 | $45,000 | $ 61,000 |
| $45,000 | $57,000 | $50,000 | $38,000 | $115,000 |

You have been asked to describe a typical household income from the neighborhood. How would you respond? Why? (You might start by considering the mean, median, and mode).

Consider the same questions for a nearby neighborhood with the household incomes: (Data Set B)

| | | | | |
|---|---|---|---|---|
| $49,000 | $57,000 | $58,000 | $46,000 | $80,000 |
| $71,000 | $74,000 | $68,000 | $52,000 | $49,000 |
| $61,000 | $59,000 | $75,000 | $43,000 | $65,000 |

In addition to the "center," what can you say about the spread of the data? If, in Data Set B, the $43,000 were replaced by $23,000 and the $80,000 by $100,000 (to form Data Set C), the mean, median and mode would not change, and the range of data would be larger than in Data Set B. Do you think that means the spread of data in Data Set C is larger?

A better way of calculating the spread is to to use the standard deviation. Here's how to find the standard deviation of a set of numbers.

1. Find the mean of the data set.
2. Subtract the mean from each of the items of data.
3. Square each of these results.
4. Add the squares.
5. Divide the sum by 1 less than the number of items of data. This is called the variance.
6. Take the square root of the variance to get the standard deviation.

Find the standard deviation first on the set {2, 12, 7}, and then on Data Sets A, B, and C.

## Activity 1

Here is another problem to try. If a population is plagued by four ailments, and the percentages having each are 80%, 75%, 70%, 70%, at least what percentage has all four?

## Activity 2

Find an example of descriptive statistics and an example of inferential statistics in a magazine or newspaper article. Cite your reference. Explain how the statistics were calculated. In the case of the inferential statistics, is the inference(s) made reasonable? What information is given to support it? What other information, if any, should have been given?

## Activity 3

Collect your own set of data. Find the mean, median, mode, variance, and standard deviation of you data set. Have your choice of data to be collected approved by your instructor.

# CHAPTER SIX - VOTING

When you think of voting probably a presidential, state or municipal election first comes to mind. Electing government officials is certainly one important type of voting. But voting also takes place for many things, such as America's most livable city, the top 25 college football teams, and best movie of the year.

In the simplest elections, there is only one candidate. For many years, such elections were all that existed in the former Soviet bloc. No idea of "fairness" or "the will of the people" comes into play when the outcome of the election is predetermined!

Today, such elections are less common, although occasionally a very popular politician will find that he or she has no opposition. In November, Americans commonly vote in elections involving only two candidates, a Republican and a Democrat. With two candidates, everyone would agree that the election should be won by the candidate receiving the larger number of votes.

### Exercise 1 - Majority and Plurality Voting

The situation becomes much more complicated when there are more than two candidates involved. If there are nine people voting, how many votes would the leading candidate have to win to obtain a majority (more than half the votes)?

Ten people?

Eleven people?

Twelve people?

n people? (Your answer will be different if n is odd than if n is even).

<u>Majority Rule</u> seems to be a fair way to decide an election. But if those voting were permitted to rank all of the candidates in order of preference, it might be possible that a candidate not winning the most votes would actually please the most people.

Let's consider an election to determine whether Andy, Blondie, or Calvin is the voters' favorite comic strip character. Nine people are eligible to vote. Each voter is permitted to vote for one character. Suppose Andy receives 5 votes and Calvin 4 votes, with Blondie receiving 0 votes. Who would win according to majority rule?

Do you think this is fair?

Suppose there were nine voters, and they filled out their preferences as follows:
Andy, Calvin, Blondie - 5 voters     Calvin, Blondie, Andy - 4 voters.

Do you think using majority rule best expresses the will of all voters, or would another candidate be a more popular choice overall? Explain.

If no candidate receives a majority, the victory can be awarded to the candidate receiving the most votes (if there is one). This is called the plurality method.

Consider an election to determine whether Axl, Bono, or Prince is the voters' favorite musical performer. Suppose each of ten voters is allowed one vote. If Axl and Bono each receive 3 votes and Prince 4 votes, who will be the winner if the plurality method is used?

What if the voters had been asked to rank their selections? Suppose the results were as follows:
Prince, Axl, Bono - 2 voters
Prince, Bono, Axl - 2 voters
Axl, Bono, Prince - 3 voters
Bono, Axl, Prince - 3 voters

Do you still think that Prince should be declared the winner? Why or why not?

When more than two candidates are involved, and voters are permitted to rank the candidates, you can see that determining the winner of an election is not so clear-cut. The following exercises will allow you to examine different voting schemes, and consider the advantages and disadvantages of each.

## Exercise 2 - Binary Voting

One voting scheme is called binary voting. Let's consider how it works in an election with three competitors. Two of the candidates compete against each other while the third gets a bye. Then the winner of the first competition goes against the one that had the bye.

Suppose there are three competitors-Dale, Earl and Florence. Voters selected their preferences as follows:
Dale, Earl, Florence - 5 voters
Dale, Florence, Earl - 2 voters
Earl, Florence, Dale - 3 voters
Earl, Dale, Florence - 3 voters
Florence, Earl, Dale - 4 voters
Florence, Dale, Earl - 2 voters

Nineteen people voted in total. If Dale and Earl were running against each other, who would win?

Now put the winner of this contest against Florence. Who would win in this case?

What would happen if Earl and Florence first compete against each other and then the winner of that contest competes against Dale?

Now see what happens if Dale and Florence first compete against each other, and then the winner competes against Earl.

Did you get the same winner in each case?

Do you think it makes a difference as to which two of the three compete against each other first?

Suppose now only 9 people had voted, and the results were as follows:
Dale, Earl, Florence - 2 voters
Dale, Florence, Earl - 1 voter
Earl, Florence, Dale - 2 voters
Earl, Dale, Florence - 1 voter
Florence, Earl, Dale - 1 voter
Florence, Dale, Earl: 2 votes

Now determine the winner in binary voting when Dale and Earl compete first, and then the winner competes against Florence.

Now have Earl and Florence compete first, with the winner then competing against Dale.

Finally have Dale and Florence compete first, with the winner going against Earl.

Now do you think the order makes a difference?

If, when using binary voting, different competitors win depending on the order in which they compete, we have what is called the agenda effect.

In the first election with Dale, Earl, and Florence, Earl always won in pairwise competition against both Earl and Florence. When one competitor always beats each of the others in pairwise competition, that competitor is called a Condorcet winner, after an 18th century French mathematician and philosopher. In the second election involving Dale, Earl and Florence, is there a Condorcet winner?

Binary voting in elections can also be used with more than three alternatives. Suppose Al, Bill, Dan and George are the candidates in an election. The ten voters filled out their slates as follows:
George, Bill, Al, Dan - 2 voters
George, Bill, Dan, Al - 2 voters
Al, Dan, George, Bill - 2 voters
Al, Dan, Bill, George - 2 voters
Dan, George, Bill, Al - 2 voters

Determine who would win in each of the six possible pairwise competitions.

Is there a Condorcet winner?

Is there a competitor who loses every possible pairwise competition in which he is involved?

A binary election with more than three competitors can be carried out using one of two methods. Let's look at how these can be used if there are four competitors (it is easy to see how these methods could be extended to an election involving more than four competitors). One method is to use a "tournament" format, where competitors are paired up, and winners of each pair then compete to see who is the overall winner. If this method is used in the previous election, the initial pairings can be assigned in three different ways. Determine who would win the election, using this method, for each of the three initial pairings. Did the same candidate always win?

Here is a second method that could be used. Have two of the candidates in direct competition, and have the other two sit out. The winner of the first round then competes against one of the people who sat out. The winner of that competition then faces the other person who sat out. The winner of that competition is the winner of the election. There are 12 different ways this could be done for a four-candidate election. Determine whether George, Al, Dan, or Bill would win for each of these twelve possibilities.

Again, did the order in which the candidates competed make a difference?

## Exercise 3 - Run-off Elections and Point Systems

If none of the candidates receives a majority of votes, run-off elections are often held. One or more candidates are eliminated and then the remaining choices are voted upon again. There are many methods that can be used to determine who is eliminated. Possibilities include eliminating the candidate(s) with the fewest first place votes or eliminating the candidate with the largest number of last place votes. If there are more than three candidates, either one candidate can be eliminated at a time, or all but two can be eliminated after the first election. Then the second and final election will involve only two competitors.

Try using several different run-off election methods on all of the elections in this chapter where the voters were able to rank the candidates.

A point system can also be used. In an election with three candidates, 2 points can be awarded for each first place vote, 1 point for each second place vote, and 0 points for each last place vote.

How would a point system work in an four candidate election?

An n candidate election?

Who would win the various elections if this system were used?

A 5-3-1 point system is often used in judging races, with anyone coming in fourth or lower not receiving any points.

How would each of the previous elections have turned out using this system?

Now comment on what you have learned.

# CHAPTER SEVEN - NUMBERS AND NUMERATION

Of all branches of mathematics, number theory is perhaps the most accessible. That is because we have worked with numbers all of our lives. One of the first things children learn is how to count. One of the first questions children are taught to answer is how old they are. If you just take 15 minutes at some time during the day to see how often you use numbers during that period, it would be much more than you probably realize. Many of the properties of numbers are very natural to you because you use numbers all of the time. The following exercises will help you learn more about some properties of numbers.

## Exercise 1 - Factoring Positive Integers

We say that a positive integer M is <u>divisible</u> by a positive integer N if when M is divided by N the remainder is 0. <u>Prime</u> numbers are positive integers that are divisible only by 1 and the number itself.

Some examples of prime numbers are 5, 13, 29, and 73.

The only even prime number is ______.

If N and M are positive integers and N is divisible by M, then M is called a <u>divisor</u> of N and we write M/N.

Positive integers that are not prime are called <u>composite</u>. Some examples of composite numbers are 15, 36, 48, and 72. On the next page, find the divisors for each of these numbers, other than 1 and the number itself, to show that each number is composite.

What if we wanted to find the divisors, or at least the number of divisors, of a large number such as 24,000? In the next exercise, we are going to find a method for determining the total number of divisors that a positive integer M has.

In order to do this, we first must be able to factor a positive integer as a product of prime numbers.

The following rules will help you to do this.

1. A number is divisible by 2 if and only if it is even.
2. A number is divisible by 4 if and only if the last two digits of the number form a two digit number which is divisible by 4.
3. A number is divisible by 10 if and only if it ends in 0.
4. A number is divisible by 5 if and only if it ends in 5 or 0.
5. A number is divisible by 3 if and only if the sum of its digits is divisible by 3.
6. A number is divisible by 9 if and only if the sum of its digits is divisible by 9.

You probably already knew rule 1. Investigate rule 2 by showing that 560 is divisible by 4 but that 822 is not.

Now investigate rules 3 and 4 by showing that 2640 is divisible by 10 and by 5, that 8765 is divisible by 5, and that 4758 is not divisible by either 5 or 10.

Problem #13 at the end of this chapter that asks you to explain why rules 5 and 6 work. There are also is rules for divisibility by 7 and 11, and you will work with these later.

Using a tree diagram may help in factoring. Here is an example that shows how to factor 5400.

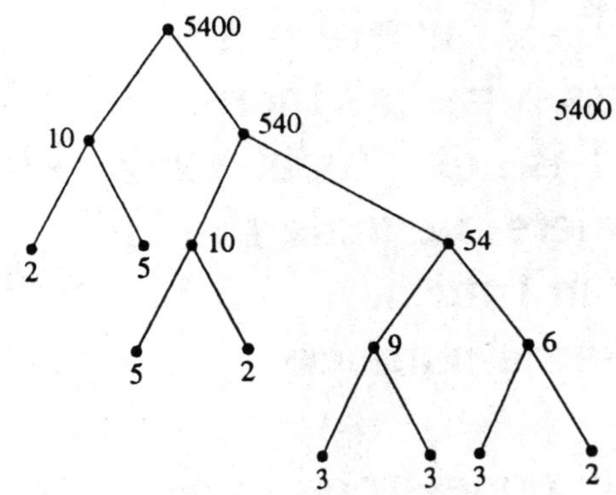


$5400 = 2^3 \cdot 3^3 \cdot 5^2$

Try using the above facts to factor the following numbers:

48 360 8100 5400 9636

No formula has been found which can be used to generate the prime numbers. One attempt was made by Mersenne, a French mathematician and monk who lived in the seventeenth century. Mersenne primes are primes of the form $2^p - 1$, where p is prime. This number cannot be prime unless p is, but there are many examples where p is prime but $2^p - 1$ is not. As of 1992, 32 such primes had been found, the largest where p equals 756839 and the resulting number is 227,832 digits in length. Supercomputers are required to find such large numbers.

Number theory has many famous and easily understood conjectures that have never been proven. One of these is Goldbach's Conjecture, which states, "Every integer greater than 2 can be written as the sum of two prime numbers."

See that the following even numbers can be written as the sum of two primes.

12

26

88

Now find some odd numbers which cannot be written as the sum of two primes.

## Exercise 2 - Counting the Number of Divisors of a Positive Integer

Now we are going to develop a method of counting how many divisors a number has without having to find them all.

First, find the number of divisors of the following numbers.
Do this by writing out all of the divisors and then counting them.

2

4

8

16

32

What relationships do you see between the numbers and how many divisors each of the numbers has?

Now try the same thing for the following numbers.

3

9

27

81

What are the similarities? How man divisors does a number of the form $p^n$ have, where p is a prime number and n is a positive integer?

The following numbers are all products of two distinct prime numbers. Count the number of divisors of each and make another conjecture.

15

21

33

51

These factor as the product of three distinct prime numbers.

30

105

How many divisors are there for a number which factors as a product of three prime numbers?

Four prime numbers?

m prime numbers?

Now count the number of divisors of the following numbers.

12

18

75

The above numbers all factor in the form $p^2q$. How many divisors does each such number have?

The following numbers factor in the form $p^2q^2$, where p and q are distinct prime numbers. Count the number of divisors of each.

36

100

The following factor in the form $p^3q^2$. Count the number of divisors of each.

72

200

Can you now make a conjecture about how many divisors a number has that can be factored in the form $p^n q^m$? Why would it be true? (Hint: Use the Multiplication Principle).

Look at all of your previous work in this exercise. Can you now make a conjecture about how many divisors a number has if you can factor it as a product of powers of prime numbers?

Try to prove your conjecture.

Find the number of divisors for 48, 360, and 8100 after first factoring the numbers. Don't find the divisors.

## Exercise 3 - GCD and LCM

The greatest common divisor of two positive integers m and n is abbreviated GCD(m, n) and is the largest integer that divides evenly into both m and n. Example: GCD(18, 12) = 6.
GCD(50, 30) = ____ GCD(36, 81) = _____.

The least common multiple of m and n is denoted by LCM(m, n) and is the smallest number into which both m and n divide evenly.
Example: LCM(6, 9) = 18. LCM(18, 27) =_____.
LCM(10, 25) = _____.

How can we find the GCD and the LCM of two numbers?

Let's start with GCD(m, n).

1. Factor m and n as a product of powers of prime numbers as discussed earlier.

2. Write down the prime numbers that are found in the factoring of both numbers. In the GCD, use as the exponent of each prime number the smaller of the two exponents that were used in the factoring of n and m.

3. Multiply together all of the prime numbers in step 2 to determine GCD(m, n).

Example: Find GCD(120, 84)

1. $120 = (2^3)(5)(3)$ $\quad 84 = (2^2)(7)(3)$

2. $2^2$, 3

3. $(2^2)(3) = 12$.

Now find the following: GCD(48, 360)

GCD(8100, 5400)

GCD(5400, 5625)

Recall the slivers problem with our friend Mr. Conehead. (page 6 #3). The denominations of the currencies were 4 and 7. These numbers are relatively prime, which means their GCD is 1; therefore, Mr. Conehead could make any purchase and receive exact change. If the denominations are not relatively prime, then a purchase can be made receiving no change or exact change if and only if the amount of the purchase is a multiple of the GCD of the two denominations. Examples: If the denominations are 4 and 6, GCD(4, 6) = 2 and only purchases which are multiples of 3 can be made exactly.

If the denominations are 9 and 12, GCD(9, 12) = 3.

If the denominations are 10 and 15, what purchases can be made exactly?

Look at problem #4 page 6 again also. How does the GCD of the sizes of containers affect what quantities of water can be measured?

Now let's consider how to find LCM(m, n).

1. Factor m and n as before.

2. Write down the prime numbers that are factors in either of the numbers. If the prime number appears in only one of the factorings, use that exponent; if the prime number appears in both factorings, use the larger exponent.

3. Multiply together the numbers in step 2 to determine LCM(m, n).

Example: Find LCM(40, 28)

1. $40 = (2^3)(5)$ $\quad\quad$ $28 = (2^2)(7)$

2. $2^3$, 5, 7

3. $(2^3)(5)(7) = 280$.

Find the following: LCM(48, 360)

LCM(5400, 8100)

LCM(5400, 5625)

Now divide (40)(28) by LCD(40, 28) and compare the answer to GCD(40, 28). Make a conjecture.

Test your conjecture on some of the values that you have already calculated.

## Exercise 4 - Other Topics Relating to Divisibility

You want to tile a floor with square tiles. Your floor measures 12 ft by 18 ft. You have tiles in the following sizes: 7 in., 9 in., 10 in., 12 in., and 15 in. Which can be used if the tiles cannot be cut?

Try to determine which of the following are true for all positive integers a, b, c, and d. Look at some examples.

1. If a/b and b/c, then a/c.

2. If a/(b + c), then a/c or a/b.

3. If a/b or a/c, then a/bc.

4. If a/b or a/c, then a/(b + c).

5. If a/b and a/c, then a/(b + c).

6. If a/b and a/c, then a/bc

7. If a is a prime number and a/bc, then a/b or a/c.

A perfect number is a number that is the sum of all its proper divisors (all of its divisors except the number itself).

Example: The proper divisors of 6 are 1, 2 and 3, and $1 + 2 + 3 = 6$. Therefore, 6 is a perfect number.

Example: The proper divisors of 8 are 1, 2 and 4. But $1 + 2 + 4 = 7$, not 8. Therefore, 8 is not a perfect number.

The next perfect number after 6 is between 20 and 30. See if you can find it. Can you find any larger ones?

It is not known if all of the perfect numbers are even, but no odd ones have yet been found. A formula to generate the even perfect numbers has been found; it involves Mersenne primes. It is not known if there are any odd perfect numbers, either. Few perfect numbers are known.

A pair of numbers is called amicable if the numbers have the following property: The divisors of the first number (excluding the number itself) add up to the second number, and the divisors of the second number add up to the first number.

What number and 284 form a pair of amicable numbers? This is the pair containing the smallest numbers.

There is a test for divisibility by 11. Add up the first, third, fifth, etc., digits of the given number. Then add up the second, fourth, sixth, etc. digits. Take the difference of these two values. If the difference is divisible by 11, so is the original number. Try this to show that 101409 is divisible by 11, but that 321429 is not.

Try some 12 digit numbers.

Can you prove that this rule is true?

## Problems for Group Solution and Discussion

The following are some problems that can be done in group discussions.

1. A group of school children went on a bus tour. Each child paid his or her own exact fare with six coins. In totaling the fares, the bus driver received a sum of $41.89. How many pennies did the driver receive?

2. There were 1000 people and 1000 doors, numbered from 1 to 1000, on one side of a very long hall. The first person went down the hall and opened every door. The second person went down the hall and closed every other door. The third person went down the hall and changed every third door, i.e. closed it if it was open and opened it if it was closed. Similarly, the fourth person changed every fourth door, the fifth person every fifth door, etc. After the 1000th person changed the 1000th door, how many doors are left open?

3. Find the units digit of the following numbers: $5^{151}$, $7^{151}$, $9^{151}$.

4. A woman with a basket of eggs finds that if she removes the eggs from the basket either 2, 3, 4, 5, or 6 at a time, there is always one egg left. However, if she removes the eggs 7 at a time, there are no eggs left. If the basket holds up to 500 eggs, how many eggs does the woman have?

5. After gathering a pile of coconuts one day, five sailors on a desert island agree to divide them evenly after a night's rest. During the night one sailor gets up, divides the nuts into five equal piles with a remainder of one, which he tosses conveniently to a nearby monkey, and secreting his pile, mixes up the others and retires. The second, third, fourth and fifth sailors all do the same

thing. In the morning, the number left less one is again divisible by 5. How many coconuts could the original pile have contained?

6. In August of 1991, some members of the PGA (Professional Golfers Association) participated in an invitational tournament at Oak Valley St. Clair Country Club. Play proceeded in the usual manner, but the scoring was done as follows: on each hole, points were awarded according to the rules below:

a. 0 points for par
b. +1 points for a birdie (one under par)
c. +3 points for an eagle (two under par)
d. +5 points for a double eagle (three under par)
e. -1 points for a bogey (one over par)
f. -3 points for a double bogey (two over par)
g. -5 points for anything worse

What scores are possible after a player has completed playing three holes? Nine holes? Eighteen holes?

7. In the problems below, each digit has been replaced by a different letter. Each letter stands for the same digit wherever it appears. Can you break the code in each problem and find the numbers?

```
G R A P E          H O C U S
- P E A R        + F O C U S
---------        -----------
A P P L E        P R E S T O
```

8. Walking in my home town some years ago now, I suddenly realized that I had been in my job for one quarter of my life. Perhaps because I was somewhat despondent at the time, I immediately asked myself how long it would be until I had been in my job for one-third of my life. Can you determine the answer?

9. A pencil, pen and notebook together cost \$1.00. A notebook costs more than two pens. A pen costs more than two pencils. Seven pencils cost more than a notebook, but six pencils cost less than a notebook. What could the prices of the three items be?

10. Three racks of bread contain 11, 6, and 7 loaves. By moving them from one rack at a time and transferring into a rack only as many loaves as are already there, rearrange them so that there are eight loaves in each rack.

11. A. In 100 A.D., Nichomachus, a Greek mathematician, noticed the pattern of sums below:

$$1^3 = 1$$
$$2^3 = 3 + 5$$
$$3^3 = 7 + 9 + 11$$
$$4^3 = 13 + 15 + 17 + 19$$

Make a conjecture concerning these sums. Try to prove your conjecture.

B. Write the sum of the cubes of the first k integers for k = 1, 2, 3, 4. Do you notice a pattern? Make a conjecture and try to prove it.

12. A printer uses a total of 1890 digits to number the pages of a book. Assuming the first page is numbered 1, how many pages are there in the book?

13. A friend of mine claims that if any number is divisible by 3, then any rearrangement of the digits in the number will give another number divisible by three. Is this so? Explain your answer. Is this true if the number 3 is replaced by the number 9?

14. Arrange the numbers 1 through 9 one in each square, so that the sum across each row, down each column, and along each diagonal is the same number. Each of the digits 1 through 9 will be used exactly once.

| | | |
|---|---|---|
| | | |
| | | |

This is called a magic square.

15. I have in mind a number that when you remove the units digit and place it at the front, you get the same result as if you had multiplied the original number by 2. Am I telling the truth?

16. Anna took 70 eggs to market and Barbara took 80. Each sold part of her eggs at 5 cents per egg and later sold the remainder at a lower price (in cents per egg); the lower price was the same for Anna and Barbara. Each received the same amount of money. What is the smallest amount that they could have received? What is the largest amount of money?

17. Explain why the following test is valid or not valid: "A natural number is divisible by 7 if and only if the difference of the number represented without its units digit and twice the units digit of the original number is divisible by 7."

18. a) Fill in the blanks with the largest fixed natural number independent of the value of N, for which you think the statement is true.

1) For any even number, $N^2$ is divisible by _____.
2) For any odd number, $N^2 - 1$ is divisible by _____.

b) Twin primes are primes that are two consecutive odd integers, such as 3, 5    9, 11    17, 19
It is not known if there are an infinite number of twin

primes. Find some more.

c) A prime triplet consists of three primes of the form N, N + 2, N + 4. For instance 3, 5, 7 form a prime triplet. Are there any more prime triplets? If there are record some. If you find none, can you prove there are no more?

19. When a certain four digit number is multiplied by 9, the product is a four digit number with the original digits in reverse order. What is the number?

20. In how many different ways can you fill in the blanks of

_ 5 _ _ 3 1 6 7 2

with the digits 4, 9 and 8, in some order so that the resulting number is divisible by 792?

21. Explain the magic trick:

Magician: "Pick any number. Double it; add 7; multiply by 5; subtract the number you started with. Remove any nonzero digit from the answer and tell me the remaining digits in any order."
Subject: "6 and 8."
Magician: "Then the digit you removed is a 3." He as correct.

## WORDSQUARES

To solve this puzzle you'll have to take several factors into account. The answer is a standard word square of eight interlocking words (four across and four down), each of which is represented by the product of the numerical values of its letters (A = 1, B = 2, etc.). The word CRAB, for examples, would be represented by the number 108 (3 x 18 x 1 x 21). To get started, look for factors that are prime numbers, especially those greater than 13. Also keep in mind that A is an "invisible" factor, since it equals 1.

| | | | | |
|---|---|---|---|---|
| | | | | = 43700 |
| | | | | = 16200 |
| | | | | = 162 |
| | | | | = 34200 |
| 6840 | 16200 | 5320 | 43700 | |

| | | | | |
|---|---|---|---|---|
| | | | | = 1008 |
| | | | | = 5700 |
| | | | | = 95 |
| | | | | = 6480 |
| 360 | 3420 | 35910 | 80 | |

| | | | | |
|---|---|---|---|---|
| | | | | = 2070 |
| | | | | = 34020 |
| | | | | = 360 |
| | | | | = 1710 |
| 18630 | 90 | 4860 | 280 | |

# CHAPTER EIGHT - GRAPH THEORY AND MAP COLORING

Graphs have applications in many different fields, including sociology, economics, archaeology, genetics, history, physics, and computer science. Graphs are a useful way to show whether there isa connection or relationship between people, circuit elements, cities, etc. What a graph is will be explored more fully after you complete the first exercise.

## Exercise 1 - The Konigsberg Bridge Problem and Eulerian Circuits

The "Father of Graph Theory" is Leonhard Euler. He solved the Konigsberg Bridge Problem in 1736 by making a graph.
Here is the problem:

The city of Konigsberg in Prussia (which was later renamed Kaliningrad in the Soviet Uniton, now once again Konigsberg in the Russian Republic), consisted of two islands linked to each other and to the banks of the Pregel River by seven bridges (see picture below).

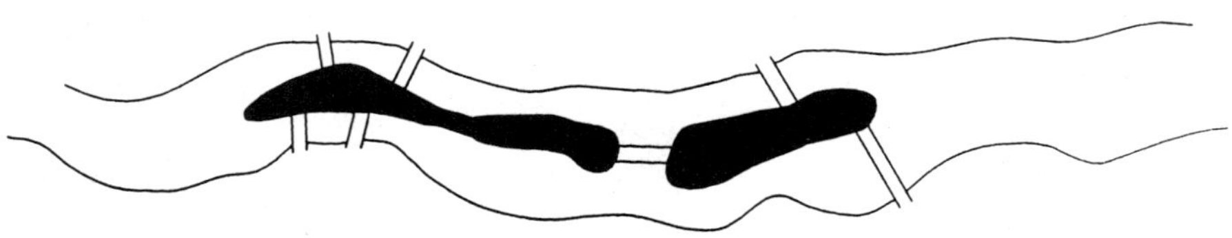

In the warm summer evenings, the citizens of Konigsberg would stroll through the city and cross over some of the bridges. The people wanted to know if it was possible to begin at any one of the four land areas, walk across each bridge exactly once, and return to the starting point.

Euler solved the problem by first turning it into a graph. His goal was to be able to start at some vertex (point), travel to another vertex via a connecting edge (line), and, by continuing this process, return to the original vertex while using each edge exactly once.

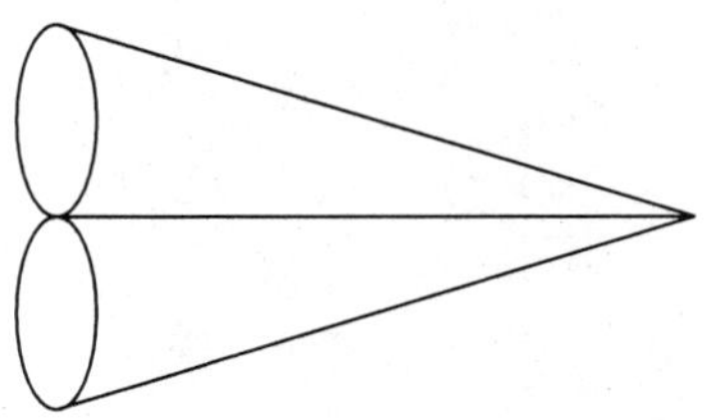

See if you can solve the problem. Can you find a way that would have satisfied the citizens of Konigsberg? If not, what seems to be the problem?

For each of the graphs below, see if you can start at any point, use each edge exactly once, and finish back at the original point. If this can be done, then the graph is said to have an Eulerian circuit. A graph must have what property to guarantee that it has an Eulerian circuit?

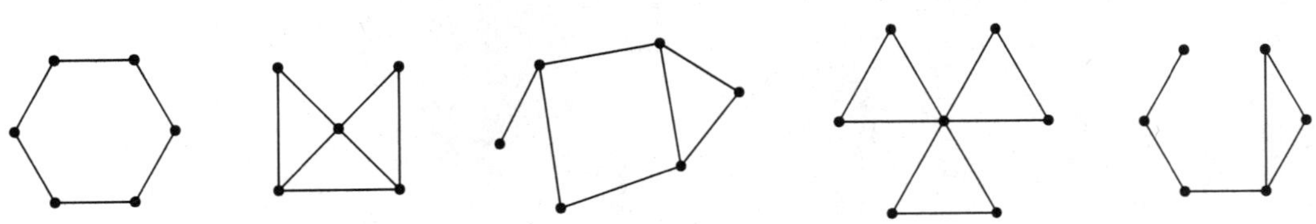

For those graphs not possessing an Eulerian circuit, see if you can start at a point, use each edge exactly once, and end up at a point other than the one you started at. What property must a graph have to satisfy this condition?

## Exercise 2 - Vertices, Edges, Degrees, and Complements

A graph is a collection of points, called vertices, and lines connecting the vertices, which are called edges. Any two vertices that are connected by an edge are called adjacent vertices.

Below are some examples of graphs.

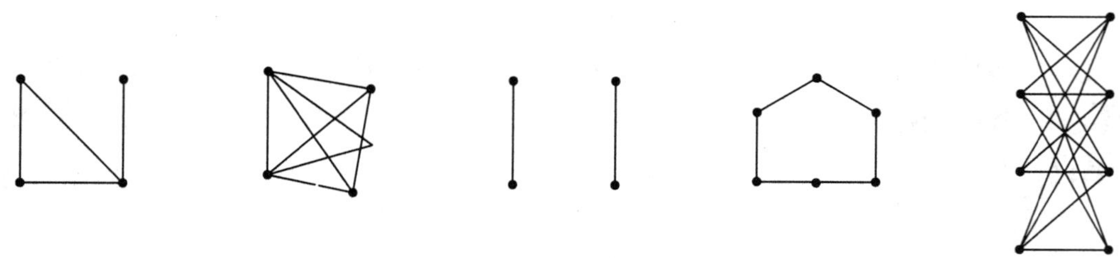

In some cases, a length and/or direction is associated with an edge.

The number of edges associated with a particular vertex is called the degree of the vertex. For each of the above graphs, calculate the degree of each vertex. Add these numbers together and compare the degree total to the total number of edges in the graph. Make a conjecture.

Can you see why your conjecture holds?

Suppose you want to design a schedule for a seven team league where each team will play only two of the other teams. Can you use a graph to show how to do this? (Hint: Represent each team with a vertex and join two vertices if the corresponding teams are to play each other).

What about a schedule where each team plays four of the other teams?

Why is it impossible for each team to play three of the other teams?

Five of the other teams?

The complement of a graph G is denoted by $\overline{G}$; $\overline{G}$ has the same vertices as G, and vertices are adjacent in $\overline{G}$ if and only if they are not adjacent in G. So if in a graph G adjacent vertices represent people who are acquainted with each other, adjacent vertices in G represent people who are not acquainted with each other. Or, strangers are adjacent in $\overline{G}$.

Find the complements of the graphs found earlier in the exercise.

If you have a party with six people or more, there are at least three mutual acquaintances (all three have met each other) or three mutual strangers (none of the three have met each other). This means that either G or $\overline{G}$ would have a "triangle" (3 mutually adjacent vertices. Draw several graphs with 6 vertices and show that for each graph G, either G or $\overline{G}$ contains a triangle.

Show by examples that if G contains 3, 4, or 5 vertices, it is possible that neither G nor $\overline{G}$ contains a triangle.

## Exercise 3 - Hamiltonian Circuits

Here is a problem that resembles the previous one, but there is a significant difference. The security man in an art gallery must check the rooms of the collection as quickly as possible. He wants to know if there is a way that he can start in one room, go into each room exactly once and return to the original room?

Here is the floor plan of the gallery.

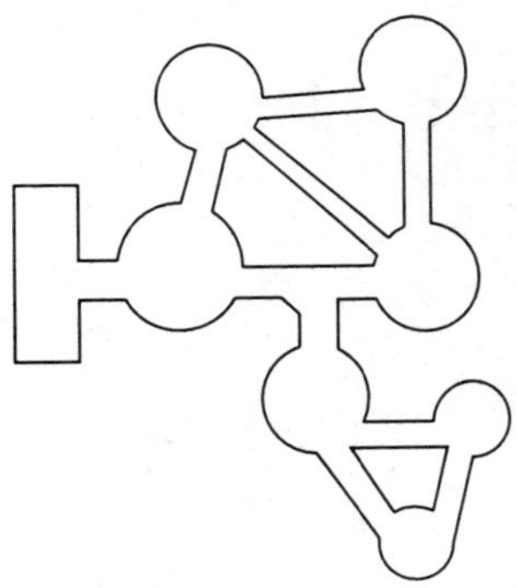

The security guard made a graph of this plan as follows. He represented each room by a vertex of the graph, and each door by an edge.

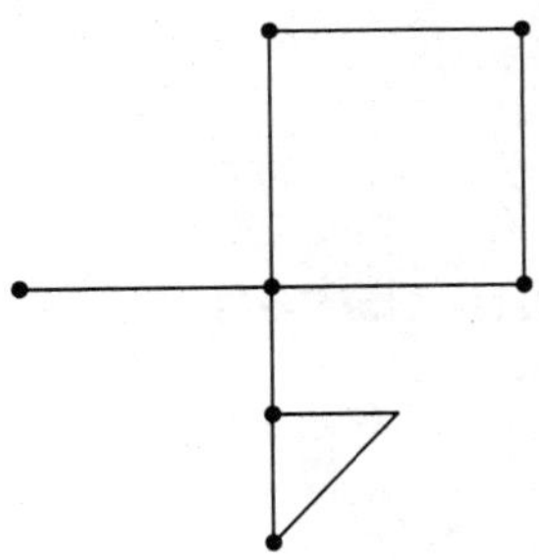

Can you start at one vertex, visit each <u>vertex</u> exactly once, and return to the original vertex? Not all of the edges need to be used. If such a circuit can be found, the graph contains a <u>Hamiltonian circuit</u>. Can you find one in the graph made by the security man?

What if room 1 (vertex 1) is omitted?

What if room 1 is omitted and one additional pathway is added?

For what types of problems would it be useful to find a Hamiltonian circuit?

Draw a few graphs and determine if each has a Hamiltonian circuit or not.

What would a Hamiltonian Path be?

Can you find any in the graphs you drew in answer to the previous question?

Unfortunately, no simple test is know to determine if a graph has a Hamiltonian circuit or not.

Hamiltonian circuits also have applications to picking up trash and sweeping or plowing streets in the most efficient way.
Can you think of some others?

What is the difference between a Hamiltonian circuit and an Eulerian circuit?

## Exercise 4 - Map Coloring

Maps are colored in such a way that no two countries containing a common border may be colored the same color (a common border consists of more than a single point). A question which has been studied for centuries is, "What is the smallest number of colors that a particular map would require?" Here are some maps that require only two colors. Why are only two colors needed?

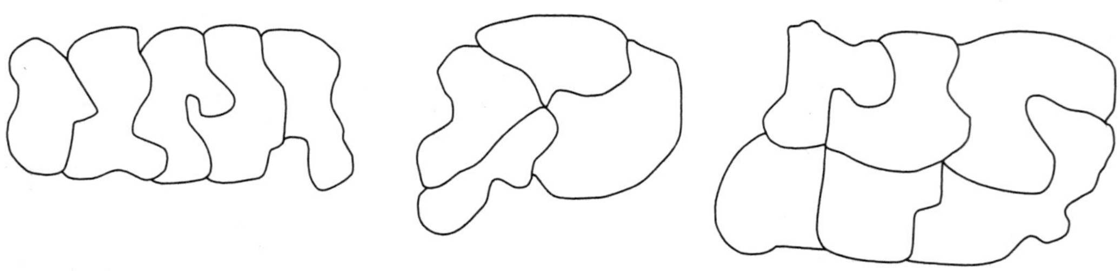

Can you draw a map that requires three colors?

What about four colors?

## Five colors?

It was conjectured long ago that four colors would be sufficient to color any map. A proof that no more than five colors would ever be required can be understood by an undergraduate mathematics major. This proof was discovered hundreds of years ago. The Four Color Theorem was finally proven in the 1970's with the help of a computer. Since the proof requires a computer, some mathematicians do not accept it.

If the surface on which the map lies is shaped like a doughnut, as many as seven colors may be required.

Instead of coloring the map itself, it is possible to draw and color a graph representing the map. Each vertex represents a country, and countries are adjacent if they share a common edge. Then, any two adjacent vertices must be colored with a different color. This can be used to find the minimum number of ways of doing something. Try using this technique on some of the maps that you colored previously.

## Problems for Group Discussion and Solution

1. My date and I attended a party where there were two other couples. Various handshakes took place. No one shook hands with himself or herself or with his or her date, and no one shook hands with the same person more than once. After all the handshakes were over, I asked each person, including my date, how many hands he or she had shaken. Each gave a different answer. How many hands did my date shake? What is the answer if there were three other couples? Four other couples?

2. Seven famous numismatists, Alban, Bela, Carla, Dinah, Ethel, Fred, and Gregor met to trade coins with each other. Each trade that took place was between only two people. After the meeting, Alban asked each of the others how many people he/she traded with. Each answered with a different number. How many people had Alban traded with?

3. June gives a party and invites six friends: two married couples and two singles. What is the maximum number of mutual strangers among the seven people at the party? If all of the women at the party know each other, what is the maximum number of mutual strangers at the party?

4. The following map is of the subway system of North Bend. Is it possible to take each route exactly once and end up back where you started? If not, can you take each route exactly once but not ended up back where you started? Try answering the above for the routes for Alphabet City and Paradise City.

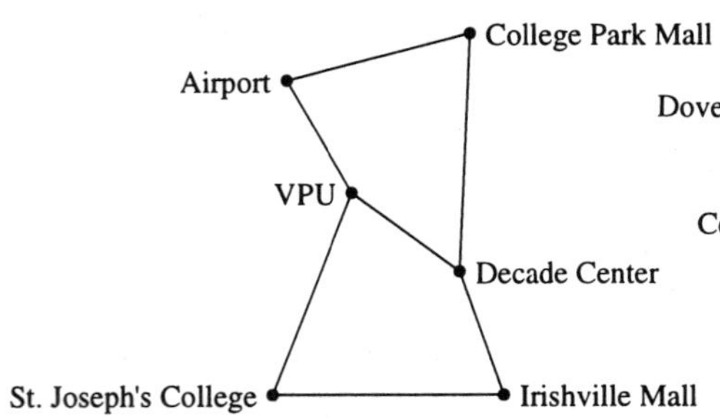


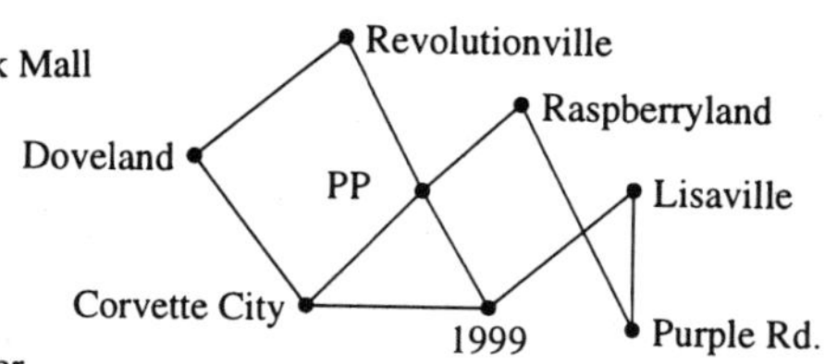


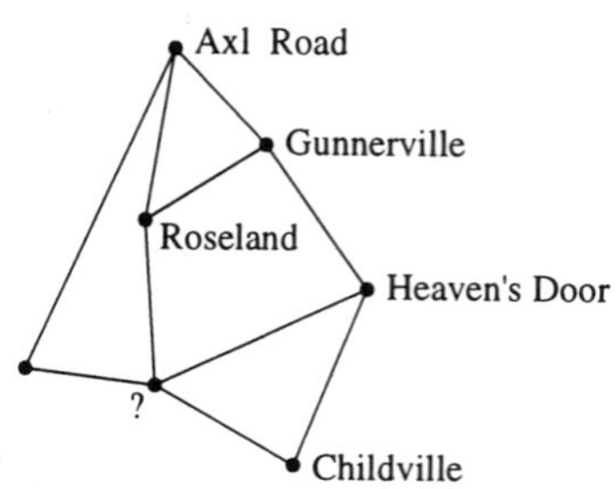

5. There are three islands and it must be possible for everyone on each of the islands to visit one another. How many bridges are needed to accomplish this? How many different ways can the islands be connected with this number of bridges? Answer the same questions if there are four islands, five islands.

6. As head of the mathematics department you have to schedule the post-year meetings. Your department offers seven different courses and the teachers that teach each course must meet at the end of the year to "wrap things up" and get ready for the next year. Each meeting will take about 50 minutes. You decide to schedule each meeting every hour on the hour the day after the students leave. This gives everyone a 10 minute break between meetings and will fit all meetings into the normal school day. One of your teachers, Joe Colorer, looks at the department schedule and sees that it is possible to have all your meetings over in much less time. Realizing that no one teaches more than three courses you think he could be right and tell him to set up a schedule that requires the least time for the meetings.

Here are the schedules for each teacher:

| | |
|---|---|
| Amadeus | Algebra I, Algebra II, Geometry |
| Barry | Algebra I, Trig, Calculus |
| Claude | Trig, Basic Math |
| Derek | Algebra I, Algebra II |
| Elvira | Algebra II, Basic Math |
| Franco | Algebra I, Trig |
| Gina | Basic Math, Functions |
| Henri | Algebra II, Trig |
| Indira | Algebra I, Geometry |
| Jonah | Algebra I, Calculus, Geometry |
| Ken | Functions, Calculus |
| Ludwig | Functions, Geometry |

7. As organizer of a musical festival, you need to organize sessions on the history (brief) of the following types of music: Rap, Alternative, Heavy Metal, Country, Roots of Rock, Impressionism (classical), and Romanticism (classical). Each session will last about an hour and half, with a fifteen minute break in between. You have twelve people that will help lead the sessions; because of their multiple areas of expertise, each person will help with more than one session. Your goals are (1) to have each session covered by all of the qualified people, and (2) by running several sessions simultaneously, to be able to complete the sessions during a five hour period so that performances can be scheduled for later in the day. Devise a schedule, using the qualifications below, that will meet your goals.

| Person | Areas of Expertise |
|---|---|
| Axl | Rap, Alternative, Heavy Metal |
| Buddy | Rap. Country, Roots |
| Claude | Country, Impressionism |
| Dolly | Rap, Alternative |
| Edie | Alternative, Impressionism |
| Frederic | Rap, Country |
| Garth | Impressionism, Romanticism |
| Kool-G | Alternative, Country |
| Latifah | Rap, Heavy Metal |
| Morrissey | Rap, Roots, Heavy Metal |
| Ozzy | Romanticism, Roots |
| Patsy | Romanticism, Heavy Metal |

Set up a schedule where all of the meetings can take place in the shortest amount of time.

8. A fixed wheel is cut into $n > 2$ distinctly labelled pie-sectors. Color each sector with one of 4 given colors. In how many ways may this be done if no two adjacent sectors are the same color?

9. The figure below shows the floor plan of a house having two exterior doors. Is it possible to enter the house through door a, travel through the house, passing through each doorway exactly once, and then exit through door i?

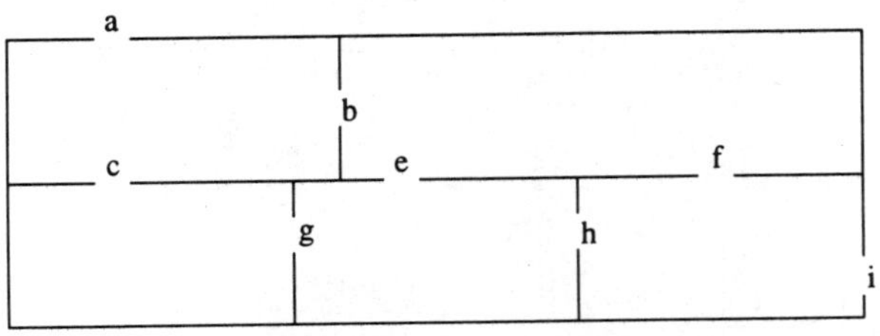


10. Suppose a map is composed of all rectangular regions (they do not all have to be the same size). The task is to color the regions in such a way that adjacent regions never share the same color. (Regions having only one point in common are not considered adjacent.) How few different colors are needed? What happens if all the regions are triangular? Square? Try some other shapes.

11. Joe gives a party and invites 5 friends: Harry, Lloyd, Oscar, Myrna, and Eloise. Harry knows Oscar and Eloise, Lloyd knows Myrna. None of the others are acquainted. a) In any group of 3 of these people, what is the probability that at least two are strangers? b) In any group of 3 of these people, what is the probability that all three are strangers?

12. What is the least number of different colors necessary to color this figure so that no two touching regions are the same color?

# CHAPTER NINE - GEOMETRY

Many of you have had a geometry course in high school. Geometry is an important discipline, and it teaches you to think logically. The geometry you studied had its roots in one of the major mathematical works of all time, Euclid's Elements, which was written more than 2000 years ago. The exercises that follow will not be like those you did in your geometry class, but rather deal with areas, volumes, surface area, perimeter, and Euler's Formula.

## Exercise 1 - Polyhedra and Euler's Formula

A polyhedron is a solid figure. The planes that form the sides of a polyhedron are its faces, the lines formed by the intersecting planes are the edges of the polyhedron, and the intersections of the edges are called vertices.

Below are some different polyhedra.

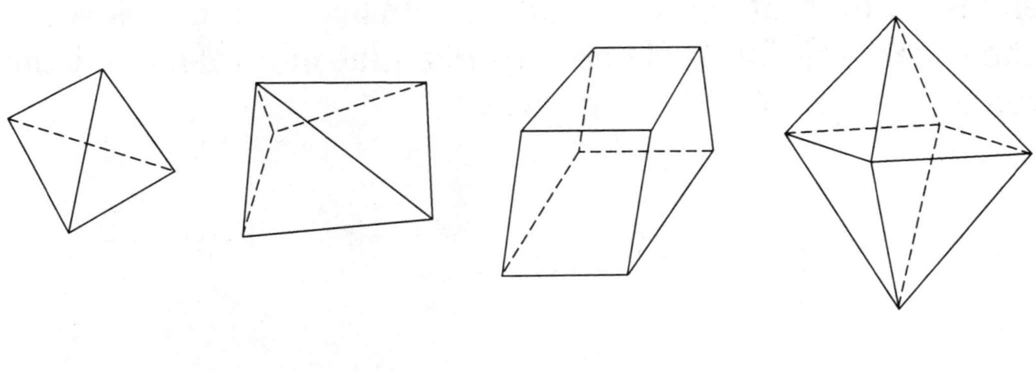

Count the number of faces, vertices and edges for each of them. There is a formula that relates the number of edges the number of faces, and the number of vertices. Can you figure out this relationship? (Hint: Try adding two of the quantities together and subtracting the third).

It is called Euler's Formula, after Leonhard Euler, one of the great mathematicians of all time and definitely the most prolific in terms of the volume of mathematics that he wrote.

A platonic solid is a polyhedron where all of the faces are the same size and shape. Some of them should be familiar to you. How many of them can you find? There are five platonic solids and they are pictured at the end of the chapter.

## Exercise 2 - Angle Measure in a Polygon

The measure of the angles in any triangle always adds up to the same number. Do you know what that is? If not, consider the following isoscells right triangle.

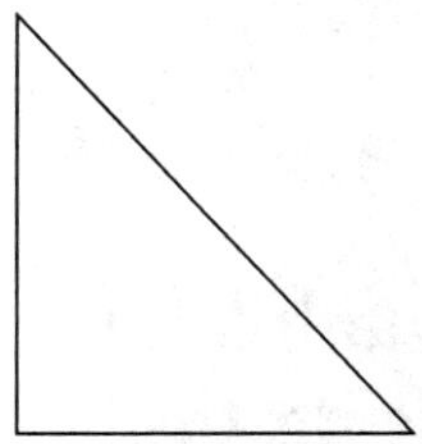

One of the angles is a right angle. What is its measure (in degrees)?

The other two angles have the same measure. What is it?

What is the total measure of the angles in this triangle?

What does the measure of the angles in a square sum to?

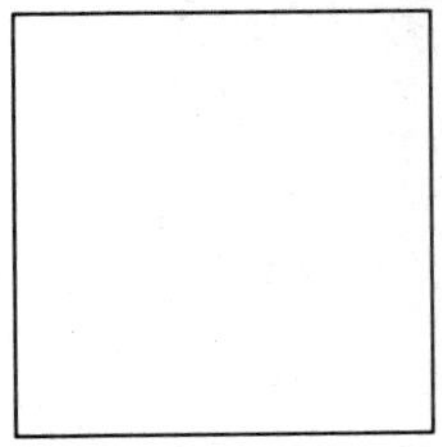

A rectangle?

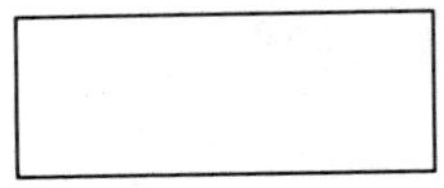

Now draw any quadrilateral (a polygon with four sides).

What is the sum of the measure of the angles of the quadrilateral? To answer this question, draw a dotted line that will divide the quadrilateral into two triangles, and use what you learned about triangles to find the total measure of the angles in a quadrilateral.

Draw a pentagon (a polygon with five sides). Divide it into triangles. What is the total measure of the angles in a pentagon?

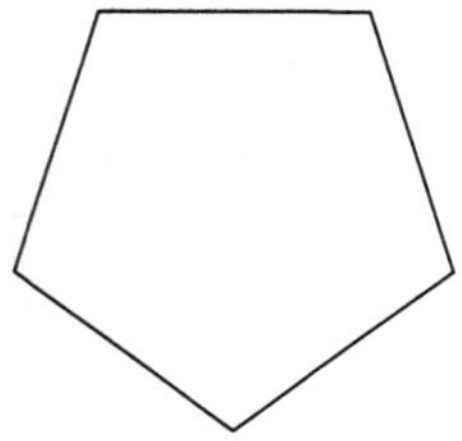

Do the same for a hexagon.

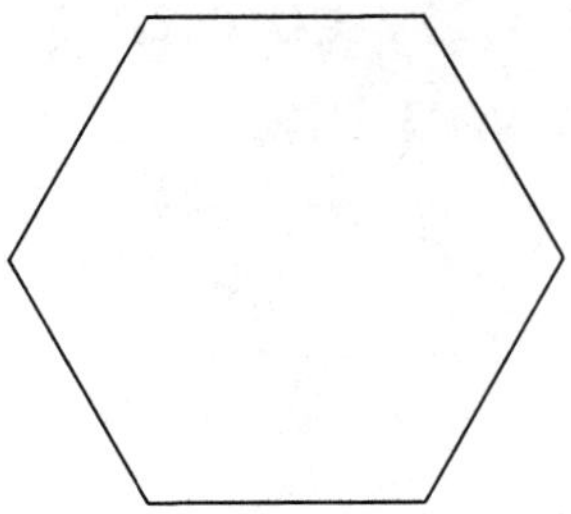

If you have a polygon with n sides, you should now be able to explain how to find the sum of the measures of the angles in this polygon. Explain how to do this in words and also by using a formula.

## Exercise 3 - Areas

Draw several rectangles on a piece of graph paper. By counting the number of blocks each rectangle encloses, can you find a formula which will give the area of a rectangle?

Now copy one of your rectangles. Cut the copy along one of the diagonals, which is the line joining opposite corners (vertices) of the rectangle. You'll have two triangles, each a right triangle. How do the two right triangles compare in size?

How do they compare in size to the original rectangle?

Can you give a formula for the area of a right triangle?

Now draw any triangle. Since we just dealt with right triangles, draw one that is not a right triangle. Make an exact copy of the triangle. Cut the copy into two pieces, each of which is a right triangle. Use these pieces and the original triangle to form a rectangle with base and height the same as that of the original triangle. What is the formula for the area of an arbitrary triangle?

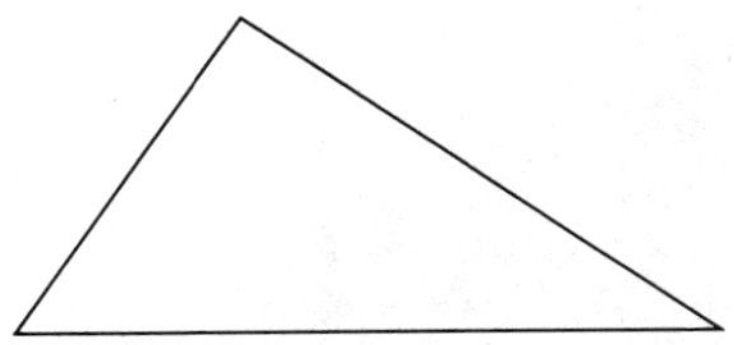

Now draw a parallelogram. By cutting a triangular piece from one end of the parallelogram and attaching it to the other end, you can form a rectangle with the same base and height as the original parallelogram. What, therefore, is the area of a parallelogram?

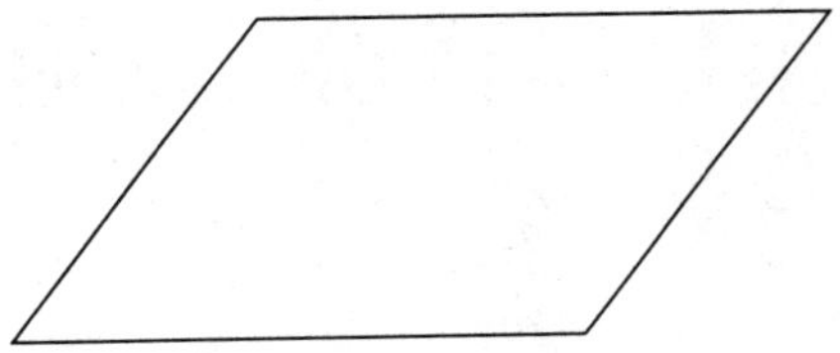

Now draw a trapezoid. By dividing the trapezoid into figures whose areas you know, can you determine the area of your trapezoid? Can you find the area of a general trapezoid in terms of the lengths of the top base, the lower base, and the height of the trapezoid?

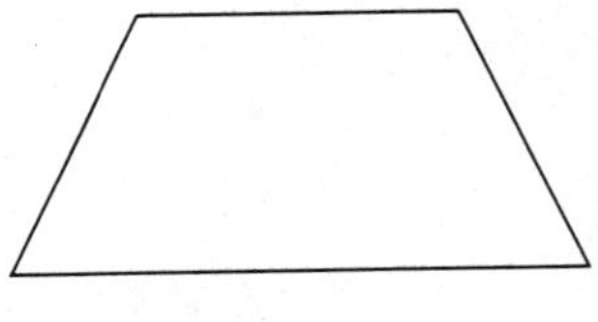

Consider the following regular pentagon. Can you determine the area of this pentagon using what we previously learned in this exercise? (Hint: Divide into triangles).

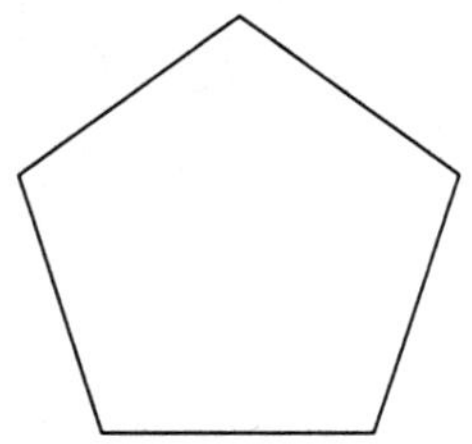

Try the regular hexagon.

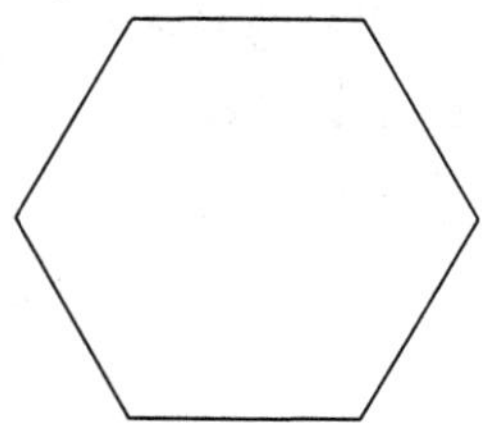

Here are several rectangles with the same perimeter-12 cm. Which one has the largest area?

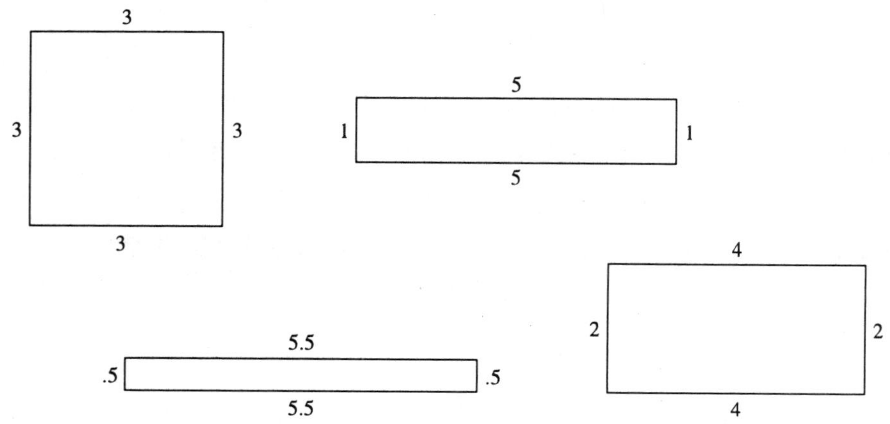

Suppose you have 60 feet of fence to enclose a rectangular garden. What should the dimensions of the garden be?

Suppose one side of the garden is against the side of your house, so only three sides of the garden require fencing. Now what should the dimensions of the garden be?

Archimedes found the area of a circle by using the "method of exhaustion." Areas of polygons, as discussed previously in this exercise, were known to the ancient Greeks more than 2500 years ago. Consider the circle of radius 1 inch. An equilateral triangle can be drawn inside the circle so that each of its vertices touch the circle. This is called an <u>inscribed</u> triangle. An equilateral triangle can also be drawn such that the circle touches each side of the triangle. This is called a <u>circumscribed</u> triangle. The area of the circle must be between these areas.

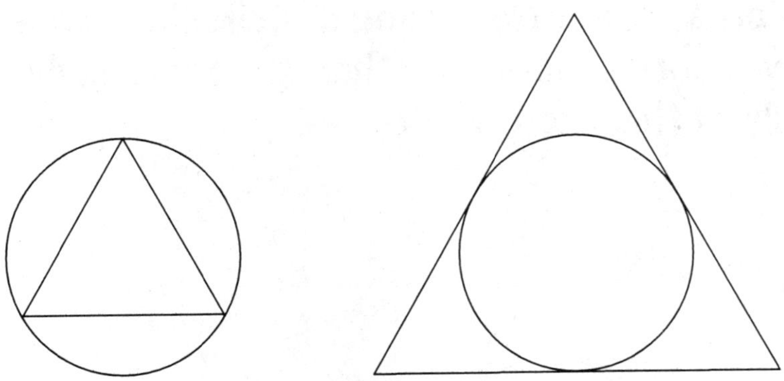

Calculate the area of each triangle. You may need to use a ruler.

A better approximation would be given if inscribed and circumscribed squares were used.

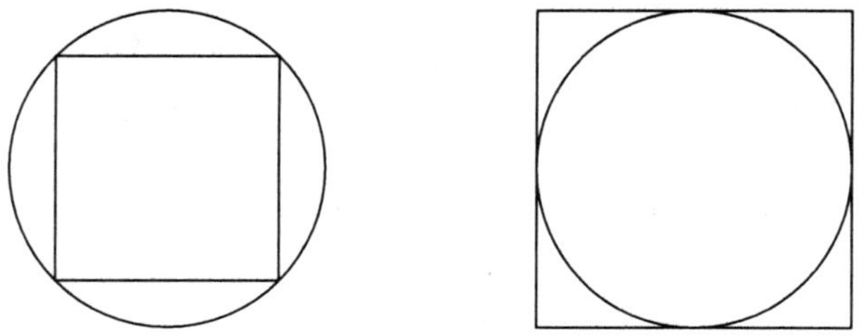

Calculate the area of the squares.

If pentagons, hexagons, and figures of more and more sides are used, the approximation for the area of the circle gets more and more exact. What is the area of a circle? Would a circular garden with boundary 60 feet have more or less area than the rectangular garden discussed previously in this exercise?

## Exercise 4 - Volumes

Use building blocks to form a rectangular box. Count the number of blocks. Can you determine the formula for the volume of such a box?

Suppose the storage area of a small truck has dimensions 6 feet by 10 feet by 8 feet. Boxes of size 2 feet by 1 foot by 1 foot are to be loaded in the truck. How many boxes could be put on the truck?

Notice that the volume of a box equals the area of one side times the length of the box. Why? Do you see that the same idea will hold even if the end is not a rectangle?

Consider the right circular cylinder pictured below. Suppose the area of the base is 6 square feet and it is 10 feet high. What is the volume of the cylinder?

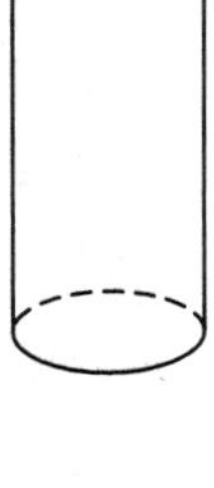

What would the volume of a metal food can be if the radius of the bottom is 1.5 inches and the can is 6 inches tall?

What if the radius of the bottom is r and the height is h?

Find the volume of the following troughs.

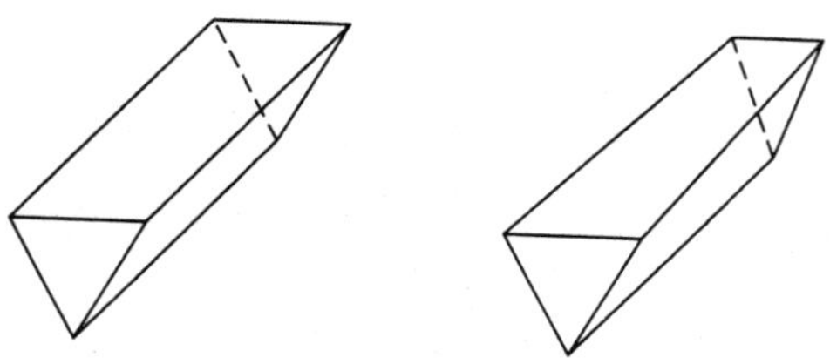

Consider the picture below. A pyramid has been put inside a rectangular box. Try to determine the volume of the pyramid from this picture. See if you can use the same idea to find the volume of a cone.

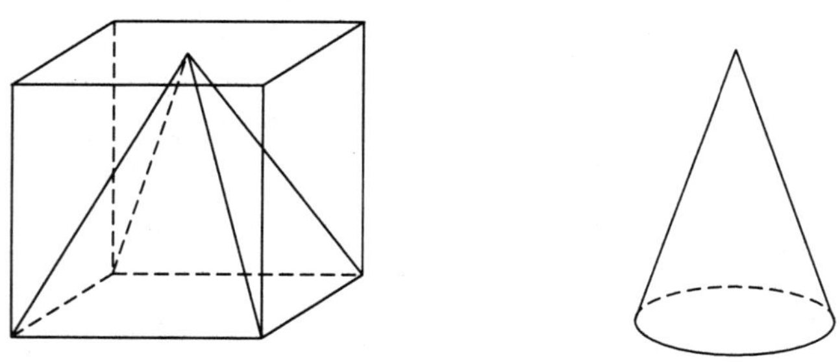

## Exercise 5 - Perimeter

The perimeter of a polygon is found by adding the lengths of each of its sides. Here are several rectangles, each of which has an area of 36 square centimeters.

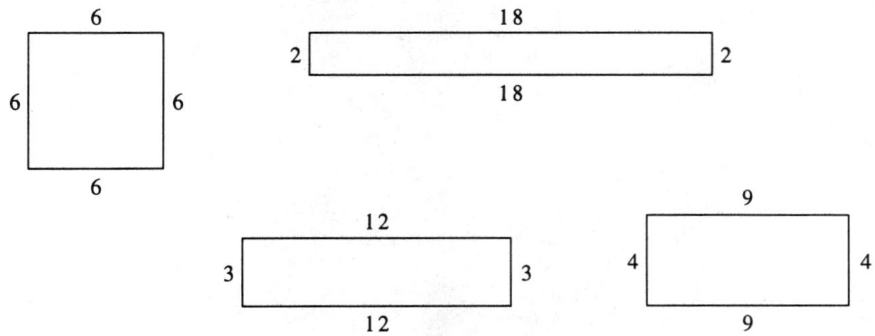


Find the perimeter of each one. Which one has the smallest perimeter?

If you were going to enclose a patch of land to make a garden, you wanted the area of the garden to be 36 square feet, and you wanted the garden to be rectangular in shape, how would you lay out the garden to minimize the amount of fence you used?

Now try answering the above question if one side of your garden is against a house and requires no fencing.

Questions involving maximizing and minimizing quantities are discussed in calculus.

Here is a rhombus with a perimeter of 24cm. Does it have an area larger or smaller than that of a square with perimeter 24 cm?

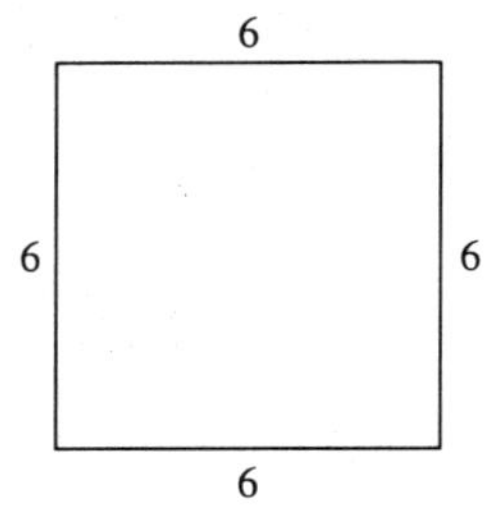


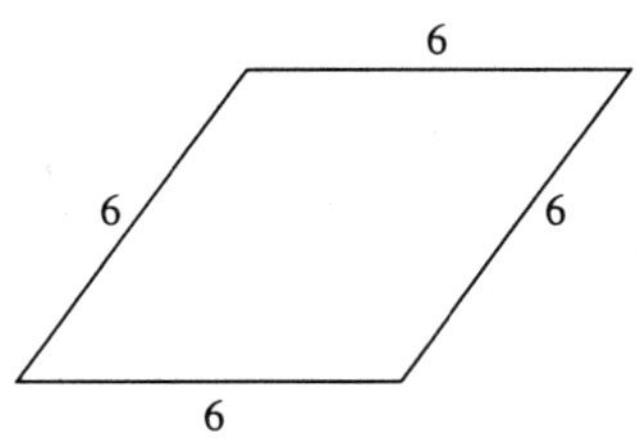


Here is another rhombus with the same perimeter. What conclusions can you make?

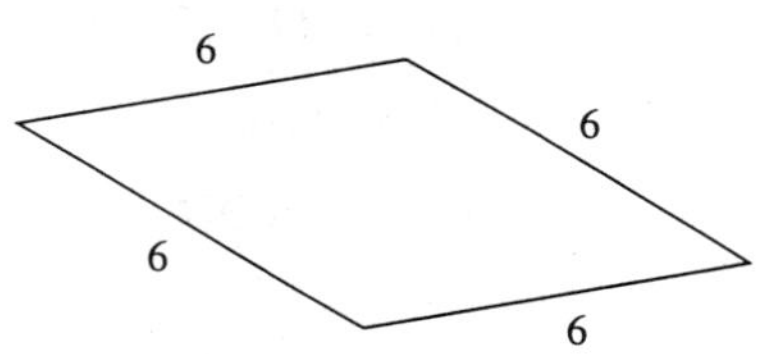


Compare the areas and perimeters of the next two figures. What can you say here?

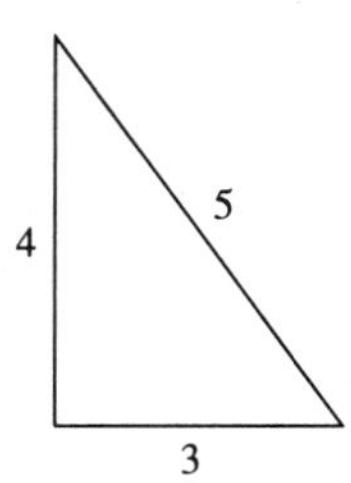


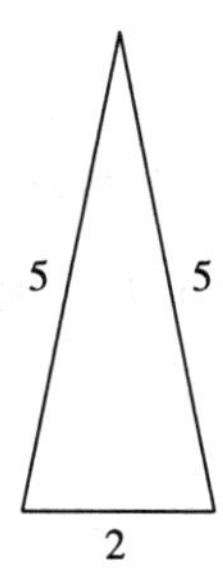

## Exercise 6 - Surface Area

If you have an aquarium or a swimming pool, you need to know what the volume of the object is in order to know how much water will be required to fill it. But if you were going to paint a room, do you need to know the volume of the room?
No, you need to know the area of the surfaces to be painted. So, an important thing to know is how to find the surface area of the room. Since we know how to find the areas of many types of polygons, and the faces of polyhedra are polygons, the surface area of many polyhedra can be found.

Try finding the surface area of the polyhedra below.

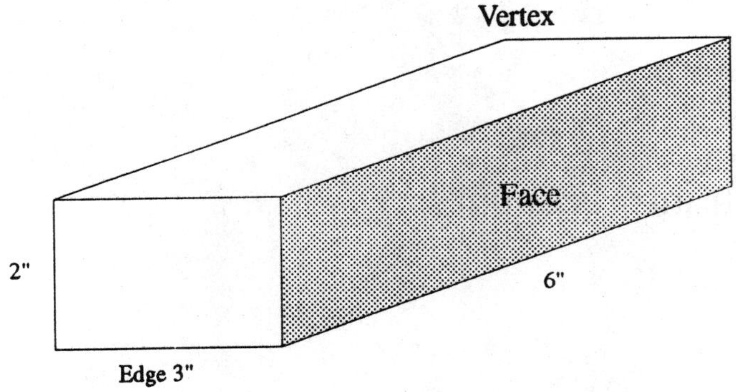


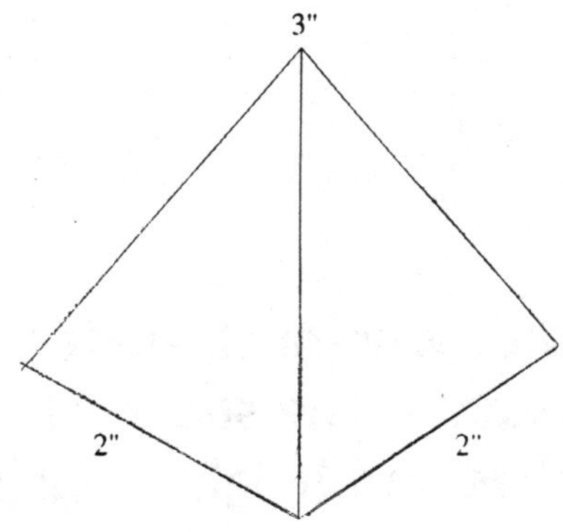


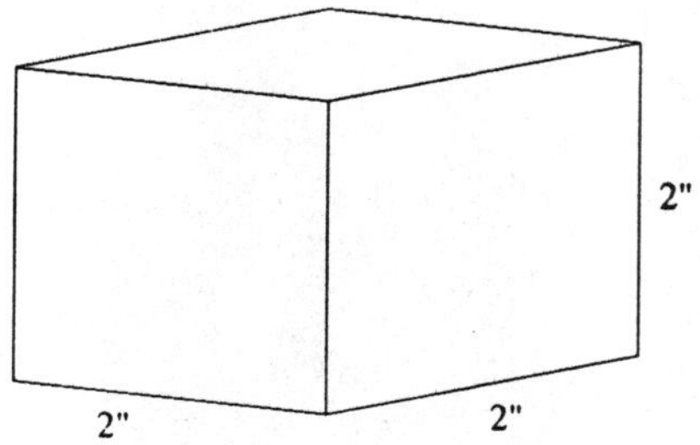

Here are the dimensions of several rooms in a house: The living room is 15' by 18', the dining room is 15' x by 12', and the family room is 20' by 15'. The ceilings are all 7' from the floor. All of the rooms are going to be painted the same color, and all of the ceilings are going to be painted white. If one gallon of paint covers 450 square feet of wall, how much paint will be required?

If the paint costs $15 per gallon and $5 per quart, how much will it cost to paint these rooms?

What factors are not being taken into account?

Now find how much material would be required to manufacture a food can in the shape of a right circular cylinder that has radius 1.5 inches and height 6 inches.

## Activity 1

Use a piece of graph paper to draw a scaled down picture of a room in your house or your dorm room or a room in your house or apartment. Include the location of windows, doors, and pieces of furniture.

Use your scale model to show two possible rearrangements of the furniture.

## Activity 2

Create a Mobius band as follows: Take a long, narrow strip of paper, twist one end 180 degrees, and then glue the ends together. Also cut another strip of paper and glue the ends together without making a twist.

Select a point in the middle of one side on each band. Put a pen on that point and draw around the band until you come back to your original starting point. What happens in each band?

Now cut both bands along the line you just drew, staying in the middle of the band. What happened?

Try the same with the Mobius band only, starting 1/3 of the way from one of the edges.

# APPENDIX 1 - LOGIC

Thinking logically is an important part not only of mathematics but also of everyday living. In this appendix we will look at some formal logic as well as some problems that simply involve logical thinking.

## Exercise 1 - Implications and their Rearrangements

You tell your friend, "If I win the lottery, I am going to buy a new car." Does that mean if you buy a new car, that you did win the lottery? If you don't win the lottery, will you not buy a new car? By rearranging parts of a statement you can form a different statement that may seem to read the same but actually has a different meaning.

An implication is a statement of the form "If _____ then _____ " having the property that that the "then" part is always true when the "if" part is true.

Consider the following implication: If month X is January, then month X has 31 days.

1. Is the implication true or false? Why?

"If month X is January" is called the hypothesis, or antecedent, and "then month X has 31 days" is called the conclusion, or consequent.

2. Create a new implication by leaving the "If" and "then" where they are but interchanging the rest of the hypothesis with the rest of the conclusion. This rearrangement of the original implication is called the converse. Is it true or false? Why?

3. Go back to the original implication. Negate the hypothesis and the conclusion. This is called the inverse of the original implication. Is this true or false?

4. Go back to the converse. Negate both parts of it. Now you have the contrapositive. Is it true or false?

Now follow the above steps to find the converse, inverse and contrapositive of the following implications (also determine if each is true or false):

a. If it is midnight, then it is dark outside.

b. If N is an even integer, then N is divisible by 2.

c. If N ends in 0, then N is divisible by 5.

d. If today is Saturday, then our class does not meet today.

e. If $x^2 = 16$, then $x = 4$.

## Exercise 2 - The Uses of "And" and "Or"

Two statements joined by the word "and" form one compound statement that is true only when both of the parts are true; otherwise, the statement is false. Determine if the following statements are true or false, and explain your reasoning.

1. February 2 is Groundhog Day and Atlanta will host the Summer Olympics in 1996.

2. Daylight Savings Time begins in April and Easter will occur in June in 1995.

3. "God Bless America" is the National Anthem of the United States and Chile is in Africa.

Joining two statements with the word "or" forms one compound statement which is false only when both parts are false.

a. "Jurassic Park" was the largest grossing movie of 1994 or $1 + 1 = 3$.

b. There are no even prime numbers or there is a largest prime number.

c. Bill Clinton is President of the United States or Elizabeth II is Queen of England.

## Exercise 3 - Logic Quiz

In the questions below, accept both statements as being correct. Then choose <u>one</u> of the answers given and explain how you obtained it.

1. It is true that (Mike is a pro football player or Mike makes more than $1,000,000 per year).
   It is false that (Mike is a pro football player).
   a. Mike makes $1,000,000 per year is true.
   b. Mike makes $1,000,000 per year is false.
   c. The given statements contradict each other.
   d. We can't determine if Mike makes more than $1,000,000 or not.

2. It is true that (Mike is a pro football player or Mike makes more than $1,000,000 per year).
   It is true that (Mike is a pro football player).

   a. Mike makes $1,000,000 per year is true.
   b. Mike makes $1,000,000 per year is false.
   c. The given statements contradict each other.
   d. We can't determine if Mike makes more than $1,000,000 or not.

3. It is true that (Mike is a pro football player and Mike makes more than $1,000,000 per year).
It is false that (Mike is a pro football player).

a. Mike makes $1,000,000 per year is true.
b. Mike makes $1,000,000 per year is false.
c. The given statements contradict each other.
d. We can't determine if Mike makes more than $1,000,000 or not.

4. It is false that (Mike is a pro football player and Mike makes more than $1,000,000 per year Mike).
It is true that (Mike is a pro football player).

a. Mike makes $1,000,000 per year is true.
b. Mike makes $1,000,000 per year is false.
c. The given statements contradict each other.
d. We can't determine if Mike makes more than $1,000,000 or not.

5. It is false that (Mike is a pro football player or Mike makes more than $1,000,000 per year).
It is true that (Mike is a pro football player).

a. Mike makes $1,000,000 per year is true.
b. Mike makes $1,000,000 per year is false.
c. The given statements contradict each other.
d. We can't determine if Mike makes more than $1,000,000 or not.

6. It is true that (If Gina will play, then the Tigers will win).
   It is true that (Gina will play).

   Will the Tigers win?
   a. Yes b. No c. Can't be sure.

7. It is true that (If Gina will play, then the Tigers will win).
   It is true that (the Tigers will win).

   Will Gina play?
   a. Yes b. No c. Can't be sure.

8. It is true that (If Gina will play, then the Tigers will win).
   It is true that (The Tigers will not win).

   Will Gina play?
   a. Yes b. No c. Can't be sure.

9. It is true that (If Gina will play, then the Tigers will win).
   It is true that (Gina will not play).

   Will the Tigers win?
   a. Yes b. No c. Can't be sure.

## Logic Project

Find a logical implication in a newspaper or magazine article. Copy the article and reference its location by giving the name and date of the publication. Identify the hypothesis and conclusion of the implication. State the inverse, converse and contrapositive of the implication. Decide whether each of these is true or false, and give reasons for your answer.

## Logical Reasoning Problems for Discussion

1. Three siblings, Alice, Bob, and Carol, truthfully reported their grades to their parents as follows:
   Alice: If I passed math, then so did Bob.
   I passed English if and only if Carol did.
   Bob: If I passed math, then so did Alice.
   Alice did not pass history.
   Carol: Either Alice passed history or I did not pass it.
   If Bob did not pass English, then neither did Alice.

2. Two women, Alice and Carol, and two men, Brian and David, are athletes. One is a swimmer, a second is a skater, a third is a gymnast, and a fourth is a tennis player. One day they were seated around a square table:
   a. The swimmer sat on Alice's left.
   b. The gymnast sat across from Brian.
   c. Carol and David sat next to each other.
   d. A woman sat on the skater's left.

   Who is the tennis player?

3. Three court cards and an ace are arranged in a row.
a. The heart is next to a spade.
b. The diamond is immediately to the left of a king or a queen.
c. The king is immediately to the left of a black card.
d. The club is not on the end.
e. The jack is one of the two middle cards.
f. The king and the queen are not adjacent.

In what order do the cards lie?

4. After a road accident, four witnesses described the hit and run driver. According to the cabby he was short and fair-haired, wearing glasses and a blue shirt. The traffic warden claimed he had dark hair and was of average height, but agreed that he was wearing glasses and a blue shirt. The truck driver thought he was a tall red-head, wearing glasses and a white shirt. The workman said he was small and bald, with a blue shirt but no glasses. It was determined that each witness described only one detail out of four correctly. Every detail was described correctly by at least one witness. What is the correct description of the driver?

5. Arrange the letters A to P in the squares of the grid according to the following clues. (The solution is unique).
a) No row or column contains either two consecutive letters of the alphabet or two vowels.
b) H is in the top row, as is B.
c) F is in the left most column.
d) I and C are in the same column.
e) G is in a corner, as are M and N.
f) P is in the third column from the left, but not in the third row.
g) E is in the same row as L.

6. Imagine a long, thin strip of paper stretched out in front of you, left to right. Imagine taking the ends in your hands and placing the right hand end on top of the left. Now press the strip flat so that it is folded in half and has a crease. Repeat the whole operation on the new strip two more times. How many creases are there? How many creases will there be if the operation is repeated 10 times in total?

7. A seven digit phone number can be detected with just 24 questions to which the answer is yes or no. Find out how.

8. Take a long thin strip of paper and hold the ends in your hands. Fold the strip in half by placing the right hand end on top of the left hand end and pressing the strip flat. Repeat this process several times. Unfold the paper and observe the creases. Some of the creases are IN and some are OUT. For example, three folds produce the sequence

OUT IN IN IN OUT OUT IN

What sequence would arise from 8 folds (if that many were possible)?

# APPENDIX 2 - EXTRA ACTIVITIES FOR STUDENTS

## Activity 1 - Scavenger Hunt

The scavenger hunt is a list of items having some mathematical connection which must be identified by the members of your group. The first group to correctly identify all items on the list will score 100%, the second group will score 95%, the third group will score 90%, etc. If some items remain unidentified or are incorrectly identified by the last full week of classes, your score will be reduced by 4 times that number.

Mathematical Scavenger Hunt:

1. Hampton Court, England.
2. The symbol for infinity.
3. The geometric shape associated with a snowflake.
4. The Mobius strip (construct one) and the Klein bottle.
5. The path which is traced by a point on the wheel of a bicycle as the bicycle moves in a straight line on a level surface. Name the path and sketch the path.
6. The origin of the word "algebra."
7. Two American presidents sharing the same birthday.
8. The number of colors needed to color Illinois, Indiana, Ohio, Kentucky, Tennessee, West Virginia, Maryland, New York and Pennsylvania so that no two contiguous states have the same color. Illustrate.
9. The Pythagorean Theorem.
10. The 11th row of Pascal's Triangle. Explain.
11. The five Platonic solids. Illustrate.
12. The number of different ways in which 3 people can be arranged around a circular table. The number of ways in which 5 people can be arranged around a circular table.
13. A Googol; how does this number compare to our national

debt?

14. The triangle suggested by a trillium. Illustrate.
15. The connection between sunflowers, piano keys and the Fibonacci numbers.
16. The quantity associated with 3.14159... . Describe its origin.
17. Hypatia - Give dates of birth and death, country of origin and contribution to mathematics.
18. Emmy Noether - Give dates of birth and death, country of origin and contribution to mathematics.
19. Tessellations.
20. Napier's Bones.
21. The Alhambra.
22. The Konigsberg Bridge Problem.
23. Tabula.
24. Nim.
25. Celtic Knots.

## Activity 2 - Article or Book Review

Reviews should range from 300 to 350 words.

The initial paragraph should be single spaced and in the following format: Author (last name, first name). Name of book. City of publication: press where it is published, date. Additional information (such as number of pages) only if you want to include it.

> Ahl, Frederick. Counting From One to Ten.
> Ithaca: Cornell University Press, 1986.

The following paragraphs, which should be double spaced, should contain:

In the first paragraph, a short description of the scope of the articles or book.

In the second (and third, fourth and fifth paragraph, depending on the number you need) a discussion (double spaced) of the most important or interesting aspects of the book or articles.

In the final paragraph the reviewer should state the audience for which such a book or articles may be useful and a general assessment of its worth and soundness.

And finally, the NAME of the author and the course in which he/she is enrolled.

Report of the National Research Council (1988):
Everybody Counts
Problem Solving FRED J. SMITH

The following list of books and articles are suggested readings for the book report. Most of these books should be available in your library. If you are unable to find book on this list or if you prefer to read another book, please check your selection with your instructor.

Abbott, Edwin: Flatland
Arem, Cynthia: Conquering Math Anxiety
Barnsley, Michael: Fractals Everywhere
Bell, E. T.: Men of Mathematics
Beckmann, Peter: A History of Pi
Boyer, Carl: A History of Mathematics
Burger, Dionys: Sphereland
Carroll, Lewis: "A Tangled Tale"
Courant and Robbins: What is Mathematics
Dantzig, Tobias: Number, The Language of Science

Davis and Hersh: The Mathematical Experience
Davis and Hersh: Descartes' Dream
Dunham, William: Journey Through Genius, The Great Theorems of Mathematics
Ernst, Bruno: The Magic Mirror of M. C. Escher
Garland, Trudi H.: Fascinating Fibonaccis
Gardner, Martin: Aha! Gotcha Gardner, Martin: Aha! Insight
Gardner, Martin: Paradoxes to Puzzle and Delight
Gardner, Martin: Time Travel and Other Mathematical Bewilderments
Gardner, Martin: Wheels, Life and Other Mathematical Amusements
Gardner, Martin: "Mathematical Fames: Euclid's Parallel Postulate and its Offspring", Scientific American 245: (Oct. 1981) p. 23-40
Ghyka, Matila: The Geometry of Art and Life
Gleick, James: Chaos
Guillen, Michael: Bridges to Infinity
Grinstein and Campbell: Women of Mathematics
Grunbaum and Shephard: Tilings and Patterns: An Introduction
Halmos, Paul R.: I Want to Be a Mathematician, an Automathography
Hoffman, Paul: Archimedes Revenge
Hofsteadter, D. R.: Godel, Escher, Bach
Hooke, Robert: How to Tell the Liars from the Statisticians
Huntley, H. E.: The Divine Proportion
Ifrah, Georges: From One to Zero
Jacobs, Harold R.: Mathematics: A Human Endeavor
Kasner and Newman: Mathematics and Imagination
Kline, Morris: Mathematics and the Search for Knowledge
Kline, Morris: Mathematics in Western Culture
Ledgard, McQuaid, and Singer: From Baker Street to Binary
Mason, John: Thinking Mathematically
Pappas, Theoni: The Joy of Mathematics

Paulos, John Allen: Innumeracy, Mathematical Illiteracy and Its Consequences

Paulos, John Allen: Beyond Innumeracy

Peter, Rozsa: Playing With Infinity

Peterson, Ivars: The Mathematical Tourist

Phillips, Hubert: My Best Puzzles in Logic and Reasoning

Polya, G.: How to Solve It

Polya, G.: Mathematical Discovery

Putnam, Hilary: Mathematics Matter and Method

Osen, Lynn: Women in Mathematics

Richards, Joan: Mathematical Visions: The Pursuit of Geometry in Victorian England

Rucker, Rudy: Infinity and The Mind

Rucker, Rudy: The Fourth Dimension, a Guided Tour of the Higher Universes

Runyon, Richard: How Numbers Lie: A Consumer's Guide to the Fine Art of Numerical Deception

Runyon, Richard: Winning with Statistics

Runyon, Richard: How to Lie With Statistics

Schattschneider, Doris: Visions of Symmetry

Schoenberg, Isaac: Mathematical Time Exposures

Smith: Mastering Mathematics

Smullyan, Raymond: Alice in Puzzleland

Steen, Lynn Arthur: On the Shoulders of Giants: New Approaches to Numeracy

Tierney, John: "Sculpting by Numbers", Science 81 2:7 (9/81)

Whitehead, A. N.: An Introduction to Mathematics

Wilder, R. L.: Evolution of Mathematical Concepts

Zappin, Leo: Uses of Infinity

April, 1974 issue of the Mathematics Teacher: Articles: "Master of Tessellations: M. C. Escher, 1898 -1972", by Ernest R. Ranucci, p. 299 "How to Draw Tessellation of the Escher Type", by Joseph L. Teeters, p. 307 "Activities: Designs with Tessellations", by Evan M. Maletsky, p. 335

## Activity 3 - Book Discussion

Instructions for Book Discussion Groups:

After you choose a book which interests you, your instructor will form discussion groups of three or four students whose books are similar. You will read your book and prepare to discuss it. Topics to be addressed in your discussion are

a. Does the book keep your interest?
b. What style does the author use?
c. What audience is the author attempting to reach?
d. Did you like the book or not? Why?

You should plan for your presentation to last about 5 minutes. As part of your presentation, you may wish to use hand outs to illustrate your comments. Since others in your discussion group will have read similar books, they may wish to ask you questions or comment in response to your presentation. The class will also be invited to take part in the discussion.

## APPENDIX 3 - Miscellaneous Problems For Discussion And Write-up

This first group of problems uses linear and/or exponential growth in their solutions.

1. The triangular numbers are integers which can represented by a triangular array as indicated below:

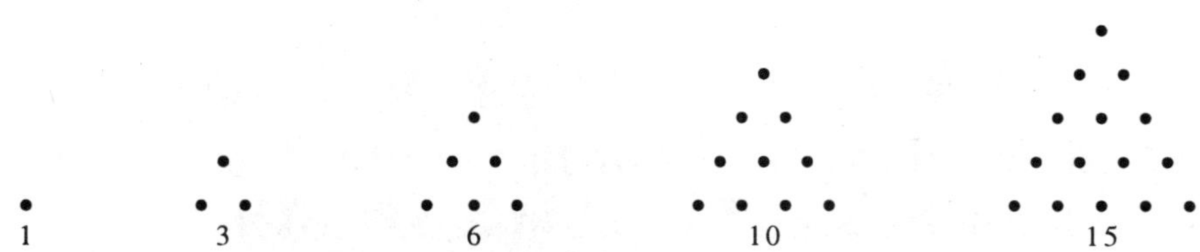


What would be the next few triangular numbers? Characterize all the triangular numbers. Could there be square numbers? Pentagonal numbers?

2. A set of dominoes is called a double-k set if it contains exactly one domino with every possible pair 0 through k including doubles (0-0, 1-1, etc.). How many dominoes are there in a double-k set?

3. Mariah C. and Barry B. both agree to work for 31 weeks and to be paid in the following manner: Mariah C is paid $10,000 the first week, $20,000 the second week, $30,000 the third week and so on to the end of the 31 week period. Barry B is paid 1 cent the first week, 2 cents the second week, 4 cents the third week, 8 cents the fourth week and so on to the end of the 31 week period. In the 31st week, which of them will have the higher salary? Which of them will make the most money over the entire 31 week period? (try to do the problem without adding up the 31 weekly salaries for each

individual). Would your answer be different if the period for payment lasted less than 31 weeks? Comment on the differences in the rate of increase of the two salaries.

4. Charlie Brown decides to initiate a chain letter. He sends a copy to each of two friends with instructions that each person should send a copy to each of two more friends within one week. If everyone who receives a letter continues the chain and no one receives more than one letter, how many letters have been sent after 15 weeks?

5. An angry ex-employee of an Admiral Motors dealership pulled the following stunt before his departure. He threw the keys to the 100 automobiles on the lot in a box after removing any identifying marks from the keys. The owner of the dealership must now match each key to the correct car. He will take a key and try it in the lock of the first car, and will continue until he finds the matching key for that car. Then he will go to the next car. What is the worst case number of trials that the owner will have to conduct to match each key with the correct car? (Assume each key is different). How would your answer change if the ex-employee had thrown in 10 more keys that didn't match any of the cars?

This next group of problems are games that students can study and or play and analyze.

1. A tennis pro is organizing a single-elimination (one loss and that player is eliminated) tournament. Initial matches are determined by drawing entrants' names, in pairs, from a hat. After the initial matches the losers are eliminated from the tournament, the winners' names are put back into the hat, and again pairs are drawn and the second round takes place. If at any time there is an odd number of names in the hat, the name left over after the

drawing of pairs remains in the hat until the next round (that player gets a "bye" for the current round).

The pro wants to know in advance how many matches will have to played to determine a single winner for any number of n people who enter the tournament. That is, what will be the total number of matches in the tournament if there are n entrants?

Answer the question if the tournament is double elimination, which means each competitor must lose twice before being eliminated.

2. Given 15 dots in a line:

. . . . . . . . . . . . . . .

Two players take turns placing an X through one of the dots. The first player to mark off a dot so that three consecutive dots are marked is the winner. Which player should win-the one who goes first or the one who goes second?

3. Consider the following game played on a 3 x 3 checkerboard. Two players alternate turns placing checkers on the board. Each player, in turn may place checkers on as many of the vacant squares as he or she wants to, provided all the checkers placed during a given turn, lie in the same horizontal or vertical column. The winner is the player who places the final checker. Who should win, the player that goes first or second?

4. This is a game for two players on a rectangular grid with a fixed number of rows and columns. Play begins in the bottom left hand square where the first player puts a mark. On his/her turn a player may put a mark into a square

directly above
or directly to the right of
or diagonally above and to the right of

the last mark made by his/her opponent. Play continues in this fashion, and the winner is the player who gets his/her mark in the upper right hand corner first. Find a way of winning which your great aunt Maud could understand and use.

5. Fifteen disks are numbered with the integers 1 through 9 inclusive. At his/her turn, each player may select one disk. The first player to select a disk such that any three of the disks selected to that point total 15 is the winner. Is it better to go first or second?

6. Two players alternately name a number from 1, 2, 3, 4, 5. A running total of all the spoken numbers is kept. The first player to say a number which brings this total to 31 is the winner. What is the best number to announce if you go first?

7. Suppose you are on a game show, and you're given a choice of three doors. Behind one door is a car; behind the others, goats. You pick a door - say, No. 1 - and the host, who knows what's behind the doors, opens another door -say, No. 3 - which has a goat. He then says to you "Do you want to pick door No. 2?" Is it to your advantage to switch your choice?

Determine the answer experimentally as follows: One student plays the contestant, another plays the host. Label three paper cups No. 1, No. 2, and No. 3. While the contestant looks away, the host randomly hides a penny under a cup by throwing a die until a 1, 2, or 3 comes up. Next, the contestant randomly points to a cup by throwing a die the same way. Then the host purposely lifts up a losing cup from the two not chosen. Last, the contestant "stays" and lifts up his original cup to see if it covers the penny. Play "not switching" 200 times and keep track of how often the contestant wins. Then test the other strategy. Play the game the same way until the last instruction, at which point the contestant instead "switches" and lifts up the cup not chosen by anyone to see if it covers the penny. Play "switching" 200 times also.

8. Five white pegs and five black pegs are laid out in a line of eleven holes as shown. I want to interchange the black and white pegs, but I am only allowed to move a peg into an adjacent empty hole or to jump one peg over another if there is an empty hole there. How can I make the interchange?

9. A magician deals 21 cards face-up into three lines of 7 cards each.

A volunteer from the audience chooses one of the cards but does not let the magician know which one. The magician asks the volunteer which line contains his/her chosen card, and then makes sure this line is placed between the other tow as all the cards are gathered into a single deck. Next the magician again deals the cards face up into three lines of 7 cards each and again asks the volunteer which line contains the chosen card. This process is repeated once more; however, this time when the volunteer tells which line the card is in, the magician announces which card was chosen and bows to great applause. Explain how the magician knew which card was chosen.

This group of problems involves weighings.

1. Eight marbles all look alike, but one is slightly heavier than the others. By using a balance scale, how can you determine the heavier one in exactly:

(a) 3 weighings? (b) 2 weighings?

2. Suppose you have a balance scale and you wish to be able to weigh any object from 1 to 63 pounds. If the weighing pan and the weight pan are separate, i.e. weights can be used on only one pan, how can you do this using only 6 weights?

Determine a set of four weights which make possible weighing loads from 1 to 40 pounds if weights can be placed on either of the pans.

3. You are given eight coins and a balance scale. Seven of the coins weigh the same, but the other one is different (you don't know if it is heavier or lighter). Find the odd coin using the fewest possible number of weighings.

A very large list of other problems is available from the publisher.